Renovating Barns, Sheds & Outbuildings

Renovating Barns, Sheds & Outbuildings

by Nick Engler

Storey Publishing

The mission of Storey Publishing is to serve our customers by
publishing practical information that encourages
personal independence in harmony with the environment.

Edited by Larry Shea and Jeanée Ledoux
Cover design by Meredith Maker
Front cover photography © Larry Lefever, Grant Heilman Photography, Inc.
Back cover photography by Giles Prett
Photography by Nick Engler
Illustrations by Mary Jane Favorite
Text design and production by Linda Watts
Production assistance by Erin Lincourt and Jennifer Jepson Smith
Indexed by Susan Olason, Indexes & Knowledge Maps

Printed in the United States by Versa Press
10 9 8 7 6 5 4

Library of Congress Cataloging-in-Publication Data
Engler, Nick.
    Renovating barns, sheds & outbuildings/by Nick Engler.
        p.   cm.
    Includes index.
    ISBN 978-1-58017-216-5
    1. Outbuildings—Maintenance and repair. 2. Barns—Conservation and restoration.
    3. Sheds—Maintenance and repair. 4. Buildings—Repair and reconstruction.
    I. Title. II. Title.
TH4961 .E56 2001
690'.89—dc21   00-026378

Contents

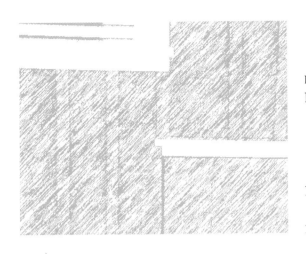

CHAPTER SEVEN *Interior walls* — *155*

Why renovate?

E very time I start out to renovate a barn or an outbuilding, someone asks me, "Why don't you just tear it down and start over?"
It's a good question. In many cases, it's easier to start from
scratch. The building techniques and materials available nowadays
enable you to put up a building quicker and more easily than you can
do a full-fledged renovation, in many cases. But quicker and easier
aren't the only considerations.

When you renovate a building, you can save three things: money, history, and architecture. The money you save, the historical value of the building, and the building's unique architecture often make a renovation much more attractive than building from scratch. So before you call in the wrecking crew, give some thought to what you might gain by letting the structure stand.

Saving history

Whether or not your farm or your home is on the Historic Register, there may be some history to your outbuildings. I've rebuilt log cabins and covered bridges that oozed history from every board. I've also helped to save structures with family and personal history. Memories — whether they belong to the general public or just you alone — are precious things.

Your barn or outbuilding may be listed on the National Register of Historic Places — or it may be worthy of listing. To determine its historical value, browse the local history section of your library. There are usually publications related to your area and its buildings. You may find early photos, deeds, and sometimes even original drawings of your structure. I once found an early volunteer fire department book containing layout drawings of every building in my town that existed when the book was made. This proved invaluable in tracing the various additions and renovations done to my property, and it helped me determine when the outbuildings were built and whether they had historical value.

1-1. This old barn has been renovated several times during its lifetime. It's an historical structure — one of the earliest in the Midwest. It's a beautiful example of early German farm architecture. And because it's on an historical register, the owner gets a tax break every time he puts money into its upkeep.

Your decision to save the history of your structure may be based purely on self-interest if the building is part of a family history. Near where I live, there is a collection of cabins and outbuildings belonging to a family who helped settle the area. They have so much pride in their heritage that they buy dilapidated buildings that have played some part in their personal pioneer history, move them to a single location, and restore them. They have a wonderful place to hold their family reunions. And they share this treasure with the general public by holding historical festivals and similar events — it's like stepping back in time to a pioneer village.

Even if your family never owned a particular structure, you may be related to the craftsman who constructed it. My three-times-great grandfather was part of a construction crew that roamed southern Ohio in the early part of the 19th century putting up public works. I've often wondered if some of the covered bridges that I've helped restore in that area weren't built by him and his friends.

1-2. The Studebaker Homestead near Dayton, Ohio, is a "pioneer village" built from homes and outbuildings that were once used by members of the Studebaker family.

1-3. The beams in this bridge may have been hewn by one of my grandsires — or yours. Sometimes that's history enough to save the structure.

Saving architecture

Many outbuildings have a unique architecture that is worth saving. In my neck of the woods, there are several round and octagonal barns. Other outbuildings work with the other homestead buildings like a matched set of salt and pepper shakers. Behind my Queen Anne Victorian home, for example, is a carriage barn that I am rebuilding. I wouldn't think of tearing it down and putting a modern garage in its place. Victorian homes and carriage barns just go together. *(See Figure 1-4.)*

Your barn may represent a unique building technique or style. *(See Figure 1-5.)* Many barns that I have worked on have contained a level of craftsmanship and style that is inspiring, and this is worth saving not just for you, but for the future. Also consider the *character* of the barn, as defined by its architecture. Could it be copied for a reasonable cost in a new structure? Could you match the details on your old building in a new one? Chances are that you can't, or that doing so will be prohibitively expensive.

1-4. This carriage barn, in great need of restoration, is located behind an old Victorian home and mirrors the home's architecture. How could you ever reproduce something with this much character?

1-5. This unique small round barn was all but rotted away when an interested group of Indiana citizens disassembled it, moved it to a new location in an historical park, and put it back to rights. Round barns were never common and this particular example, with a huge cistern in the upper story for watering livestock, makes it one of a kind.

A barn architecture primer

What architectural style does a barn represent? Unlike houses, which are often labeled for the period in which they are built, barns are classified by the origins of their *design* and their *purpose*. The clues to a barn's architecture are found in both its overall layout and the details of the structure.

It's important that you understand the architecture before you begin a restoration. Not only does this help you preserve the character of the building, it helps you plan what you must do to adapt the structure to your own purpose.

Design

Barn design evolved quickly on the North American continent. As the frontier gave way to settlement, pioneers from various ethnic backgrounds met in the New World and traded ideas about rural architecture and construction.

The earliest barns were *crib barns* — enclosures made by stacking logs horizontally, as for a log cabin. Most had just a single enclosure, but as farmers needed more space, they often added a second crib. Occasionally, you see an old log crib barn with three or four cribs.

Crib barns

As the frontier disappeared, agriculture became the base of most local economies, and prosperous farmers needed increasingly larger structures. The timber-framed post-and-beam barn began to replace crib barns for the simple reason that a larger space could be enclosed in less time using fewer materials.

Some of the post-and-beam structures were adaptations of the older crib layout. The most common of these, the *transverse frame barn,* is descended from a two-crib barn with an aisle in between the cribs. It usually has doors on the gable ends, an aisle down the center of the barn, and cribs or stalls on either side of the aisle.

Transverse frame barn

(continued on next page)

A barn architecture primer — *continued*

Settlers were quick to shed the privations of the pioneer days and, as soon as they could, they built sophisticated barns whose designs were rooted in Old World traditions. One of the earliest post-and-beam barns built in America was an English *three-bay barn*. The frame encloses three separate spaces, and the middle space — called the runway — is usually open. The spaces on either side usually have animal stalls or equipment storage below and hay lofts above.

Several types of barns were derived from this common English design. The *bank barn* was a two-story barn. The upper story was built much like that of the three-bay barn, while the lower story was excavated from the side of a hill. A stone foundation enclosed the lower story and supported the wooden upper story. You could enter the barn from either story. Typically, the farmer kept animals in the lower story and stored equipment and hay in the top story. Occasionally, in flat country with no hillsides, a farmer built a tall stone foundation and then hauled in dirt to make a ramp to the second story. This was called a *raised barn*. If he omitted the dirt ramp, it was a *foundation barn*.

Three-bay barn

Bank barn

Raised barn

German immigrants preferred to build *forebay barns.* The walls of these structures were either made from stone or framed with posts and beams. In both cases, the barns were built with an overhang or *forebay* along one side or one end. This provided protection for livestock that gathered under the forebay or farm equipment that was parked beneath it. Like English barns, these structures were built on both level ground and hillsides. *Bank forebay* and *raised forebay* barns are common.

There are two major types of German forebay barns. The *standard Pennsylvania barn* has a symmetrical gable — both sides of the roof are the same. The *Sweitzer barn* has an asymmetrical roof. The side of the roof over the forebay is longer and the forebay eaves are lower than the eaves on the other side. Within these two categories, there are many variations, all having to do with the appearance of the forebay. In a *posted forebay barn,* the forebay is supported by posts. If the forebay is framed in and enclosed, it's an *enclosed forebay barn.* If the barn has forebays on two or more sides, it's a *multiple-overhang barn.*

(continued on next page)

Standard Pennsylvania forebay barn

Sweitzer forebay barn

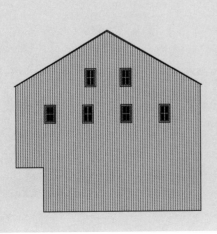

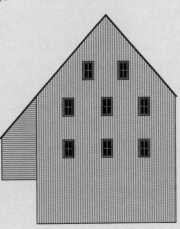

When you look at a forebay barn from the gable end, the standard Pennsylvania barn roof is symmetrical but the Sweitzer barn roof is not.

A barn architecture primer — *continued*

Another German contribution to rural architecture is the *extended barn* or, more commonly, the *three-gable barn*. This structure has an L-shaped or T-shaped layout. If the extension is on the side with the forebay, it is called a *frontshed extended barn*. If it's on the opposite side of the forebay, it's called an *outshed extended barn*. Often, an outshed extension is confused with a *porch*. But porches are really enclosed entranceways that were added to the barn after it was built. The ridge of an extension is even with the ridge of the main barn, while the ridge of a porch is always lower than the main ridge.

The French tended to make their barns longer than the barns of other ethnic groups. The *long barn* is a post-and-beam structure that may have as many as eight bays, compared with the three in a typical English or German barn. If they needed a smaller outbuilding, the French built a square structure with a gabled roof over the hay loft and a shed roof over the granary. This is called an *Acadian barn*. The French pioneers who migrated south to Louisiana developed a small, square barn with a recessed opening in the gable end. This is a *Cajun barn*.

Extended forebay barn

Long barn

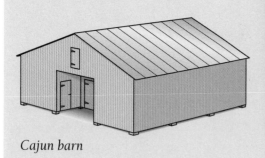
Cajun barn

Acadian barn

Dutch settlers contributed two useful rural structures. The standard *Dutch barn* looks like a saltbox house with a steep gabled roof and low sides. As the Dutch began to move west, they extended this design to make a barn whose gabled ends were much wider than the sides were long. This became known as the *Michigan Dutch barn*.

People of many other nationalities contributed to American rural architecture, of course — Swedes, Finns, Czechs, and Spaniards, just to name a few. There were also a few religious groups, such as the Mormons and the Amish, that came up with their own barn designs. But these were not as widespread as those designs contributed by the English, Germans, French, and Dutch.

As settlers began to trade ideas and agriculture developed, new barn designs evolved. For example, farmers combined the traditional English three-bay barn with the German extended forebay and got the *three-end barn* — an L-shaped or T-shaped barn without an overhang.

(continued on next page)

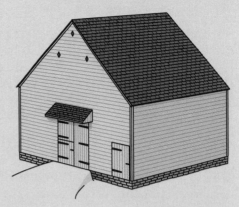

Dutch barn

Michigan Dutch barn

Three-end barn

A barn architecture primer — *continued*

Round and *octagonal* barns became popular in the later half of the 19th century as the forest gave way to farms and the large timbers necessary for post-and-beam construction became scarce. Farmers adopted this unique geometry because it allowed them to enclose more space with fewer materials. However, it was only a temporary solution to the depletion of traditional building materials. As lumber mills proliferated, they sawed up smaller second-growth trees to create dimensional lumber. As this lumber became more common and less expensive, farmers began to frame barns by nailing together wooden frames and trusses. These frame-construction barns became known as *feeder barns*. Outwardly, they look much like a three-bay or a transverse frame barn. Inside, they are framed with sawed lumber. Pegged mortise-and-tenon joints gave way to nails and gussets.

Octagonal barn

In the first half of the 20th century, farms became more mechanized and the need for equipment sheds grew. At the same time, the proliferation of electric and telephone utilities created the telephone pole — a round post that was chemically treated so it could be set directly in the ground. These two developments gave rise to the *pole barn* — a one-story barn with wide gables and a shallow roof pitch. Early pole barns were sided with vertical boards, much like a traditional barn. Later structures are clad in metal sheathing.

Feeder barn

Purpose

More so than in any other structure, form follows function in a barn. For this reason, barn architecture depends heavily on the purpose of the structure.

A *dairy barn* is a huge, two-story, transverse frame barn designed to accommodate milk cows. The cows rest in the lower story, while the upper story holds hay. The lower story is often ringed with windows to provide sufficient light for milking and other dairy operations.

In mountainous areas, where farms were small and agriculture less prosperous, barns tended to be smaller. *Meadow barns* sometimes have only one or two bays. As their name implies, they are usually built in a meadow or on another level piece of ground.

(continued on next page)

Pole barn

Meadow barn

Dairy barn

A barn architecture primer — *continued*

In areas where farmers grew tobacco, they developed *tobacco barns* to cure the crop. These were well-ventilated structures with ventilators along the ridge of the roof and gaps between the boards on the sides and ends. This let air pass freely through the structure, drying the tobacco leaves.

A *horse barn* is a small transverse frame or three-bay barn with horse stalls and hay lofts. Wagons were often parked in the runway between the stalls. The urban reincarnation of this structure is the *carriage barn,* which held horses, hay, and a carriage. As the horse and buggy gave way to the automobile, carriage barns evolved to become *garages.*

(continued on page 14)

Tobacco barn

Horse barn

Carriage barn

Garage

A barn architecture primer — *continued*

Other rural structures

While barns are the most prominent type of outbuilding on a farm, they are not the only type. Farmsteads often included many buildings, each designed for a specific purpose. *Sheds* were built for storage or workshops; *spring houses* stored perishable foods. *Smokehouses* were used to smoke meats, and *bake ovens* were used for baking bread. *Mills* of various types processed grain, sorghum, cider apples, and other crops that needed to be crushed. *Sugarhouses* helped evaporate liquids to make syrup and molasses. *Granaries*, *corn cribs*, and *silos* stored grain.

I haven't room in this book to discuss all of these different structures, and there is no need to do so. The design and construction methods used to make these smaller structures are exactly the same as those used to build barns. If you put a new roof on your spring house, you must deal with the same architectural concerns that you face when putting a new roof on your barn. You just won't have as much roof to cover.

Mill

Spring house

Shed

Saving money

The most common reason for restoring an outbuilding is that it just costs so much to put up a new one. More to the point, if you tear down the old building and put up a new one, you have to absorb the demolition and construction expenses all at once. Even if restoring an older building costs you as much (or more) than building a new one, you can perform the restoration a little at a time, stretching the cost out over a longer period. Often, this prevents you from having to borrow money and saves the interest you would have paid on the loan.

Saving and restoring an existing building might cost you more in labor than building new, but it will save you big time in material costs. As long as the structure isn't too dilapidated, a restoration will require only a fraction of the materials that would be necessary to create a new building. And if you are providing most of the labor as "sweat equity," the cash outlay for a restored building will be far less than that for a new one.

Of course, some barns really aren't worth saving, and sometimes the reason isn't immediately apparent. I was once called out to evaluate a big three-end barn. Once there, I discovered that it was an early 20th-century feeder barn, all framed in rough, green oak and nailed together. The former owner had covered the roof in asphalt shingles — a bad choice for barns because they blow off so easily. As he lost shingles, he hadn't kept up with roofing repairs, so the weather had begun to rot the wooden frame. Much of the barn was sound, but two corner posts and much of the top plate needed to be replaced. *(See Figure 1-6.)*

I told the folks who had recently acquired the farm that they should raze the barn and build new. They were surprised — the barn wasn't falling down. Why couldn't they simply replace the rotted frame members? The problem, I told them, had more to do with the construction than the rot. As green oak dries, it grips the nails like a vise and makes them almost impossible to remove. If the barn were a post-and-beam structure and all I had to do was knock out a few pegs to remove a rotted member, it would have been feasible to save the structure.

1-6. It's a shame, but this feeder barn can't be saved economically. Its construction method makes it extraordinarily time consuming to remove and replace rotted and broken members. The cost of labor would be prohibitive.

The costs of renovation

To make an informed decision as to whether or not to restore an outbuilding, you need to know the costs of a renovation, both in terms of money and time invested. To help you figure that out, I put together this chart. It lists typical tasks in a renovation and the costs associated with it. I've included the costs of hiring it done and the cost of doing it yourself — including the time you should plan on spending. Costs of rental and labor will vary regionally.

FOUNDATION

TASK	HIRING IT DONE		DOING IT YOURSELF	
	JOB COST	LABOR	MATERIAL	TIME INVESTED[1]
Foundation excavation	$150 for equipment delivery $.14 per square foot 5000 square feet per 8 hours	$700 per day	$350 per day for backhoe rental	You can excavate 50 square feet every 10 minutes.
Basement excavation	$150 for equipment delivery $.61 per square foot 1400 square feet per 8 hours	$850 per day	$350 per day for backhoe rental	You can excavate 43 square feet every 15 minutes.
Footing excavation	$150 for equipment delivery $.74 per square foot 1000 square feet per 8 hours	$750 per day	$350 per day for backhoe rental	You can excavate 30 square feet every 15 minutes.
Pouring a concrete foundation wall	$7 per square foot	$2 per square foot	$1.50 per square foot for concrete	You can form a 16' x 24' foundation wall in one week and pour it in one day.
Building a block foundation wall	$10 per square foot	$2.50 per square foot	$5.50 per square foot for block and mortar	You can lay a 20-foot section 5 feet high in one day.
Rebuilding a stone foundation	Using salvaged stone	$800 per day	$45 per linear foot using salvaged stone	You can repair about 200 square feet of stone foundation wall per day.
Jacking up a structure	$4000 to raise a 20' x 40' carriage barn	$3200	$800	With one helper, about 2 weeks

FLOORS

TASK	HIRING IT DONE		DOING IT YOURSELF	
	JOB COST	LABOR	MATERIAL	TIME INVESTED[1]
Excavating a floor for concrete slab	$1150 for equipment delivery 5000 square feet per 8 hours	$1000 per day	$350 per day for backhoe rental	You can excavate 150 square feet every 15 minutes.
Pouring a concrete floor	$400 for a 16' x 24' floor	$1000 per day	$400 for a 16' x 24' floor	You can pour a 16' x 24' slab in one day.
Replacing 16' x 24' floor joist system	$600 for floor joist materials	$1000 per day	$200–$600 for materials to fix a 16' x 24' wood floor	You and a helper can repair and replace a 16' x 24' floor system in one day.
Installing a reclaimed pine floor	About $4200 for material and installation for a 16' x 24' floor	$750	$2500 in materials; no cost if you reuse flooring	You can sand a 16' x 24' floor in one day.

[1] Assumes an experienced worker

FLOORS — continued

TASK	HIRING IT DONE		DOING IT YOURSELF	
	JOB COST	LABOR	MATERIAL	TIME INVESTED[2]
Sand, edge, fill, and finish wood floor	About $1150 for a 16' x 24' floor, includes materials	$750	$250 per week to rent sander, $50 in paper, $150 in varnish	Sanding a 16' x 24' floor should take at least one day; finishing, 3–4 days.

STRUCTURE AND FRAMING

TASK	HIRING IT DONE		DOING IT YOURSELF	
	JOB COST	LABOR	MATERIAL	TIME INVESTED[2]
Replacing a post	About $800 to have a 12" x 12" x 12' post replaced	$300	$500 for a 12" x 12" x 12' oak beam[3]; $40 each for steel jackposts	You and a helper can replace a post or two per day.
Replacing member in a post-and-beam structure	About $400 to have a 6" x 8" x 10' oak beam replaced	$250	$150 for a 6" x 8" x 10' oak beam	You and a helper can replace a beam or two per day.
Bent replacement	About $9000 to have a 20' x 20' oak post-and-beam single-story bent with king post truss system replaced	$6000	$3000 in materials[4] for a 20' x 20' oak post-and-beam single-story bent with king post truss system	This would take one day to assemble and one day to put up, with 4–6 helpers.
Sistering member in frame structure	About $150 to sister (install a new member alongside the old) a 2" x 10" x 16' floor joist	$120	$30 for a 2" x 10" x 16' floor joist and a floor jack	This would take 2–8 hours, depending on the location of the member.
Building new frame structure	About $12,000 to build a 1½-car garage	$8500	$4500 for a garage "kit" from a home center; additional $750 for slab foundation	With 2 helpers, it would take 3–4 weeks to pour the slab and put up the structure.

SIDING AND PAINTING

TASK	HIRING IT DONE		DOING IT YOURSELF	
	JOB COST	LABOR	MATERIAL	TIME INVESTED[2]
Installing new pine board-and-batten siding	About $1400 for a 16' x 24' wall	$400	$950 for a 16' x 24' wall	You can do this in one day with a helper.
Installing vinyl siding	$1.75 per foot installed; About $700 for a 16' x 24' wall	$300	$350	You can do this in one day with a helper.
Painting new wood siding	About $600 for a 16' x 24' wall	$500	$500 per week for pressure washer rental; hand scrape and sand, about $150 for paint	Prep a 16' x 24' wall in one day, and paint the next.
Painting aluminum or vinyl siding	About $500 for a 16' x 24' wall	$300	$500 per week for pressure washer rental; $200 per week for airless rental; about $100 for paint	You can pressure wash an entire structure in one day and probably spray paint the entire thing in one day with a helper.

[2] Assumes an experienced worker
[3] Select Structural Oak

(continued on next page)

The costs of renovation — *continued*

INTERIOR WALLS				
TASK	**HIRING IT DONE**		**DOING IT YOURSELF**	
	JOB COST	**LABOR**	**MATERIAL**	**TIME INVESTED**[4]
Installing wall frame in a post-and-beam structure	About $500 to frame a 10' x 20' section of a post-and-beam barn wall	$420	$80	You can do this in 1½–2 days.
Covering an 8' x 8' wall with sheetrock	$100	$60	$30	You can do two of these per day.
Covering an 8' x 8' wall with paneling	About $75	$50	$25	You can do this in about 2 hours.

ROOFING				
TASK	**HIRING IT DONE**		**DOING IT YOURSELF**	
	JOB COST	**LABOR**	**MATERIAL**	**TIME INVESTED**[4]
Installing asphalt shingles over existing 20' x 20' roof section	$250	$120	$100	You can do this in one day.
Removing material on a 20' x 20' roof section and installing asphalt shingles	$300	$200	$100	You and whoever you talked into doing the dirtiest job on earth can probably do this in one day.
Removing material on a 20' x 20' roof section and installing galvanized ribbed roofing	$600	$350	$250	You can do this in 8 hours.
Removing material on a 20' x 20' roof section and installing 24" shake shingles	$1100	$450	$500	You can do this in 8–16 hours with 2 helpers.
Installing new Vermont slate roof on a 20' x 20' section	$2700	$700	$2000	You can do this in 8–16 hours with 2 experienced helpers.
Installing Spanish tile on a 20' x 20' roof	$3000	$600	$2500	You can do this in 8–16 hours with 2 experienced helpers.

[4] Assumes an experienced worker

DOORS AND WINDOWS

TASK	HIRING IT DONE		DOING IT YOURSELF	
	JOB COST	LABOR	MATERIAL	TIME INVESTED[5]
Installing a new steel entry door	$250	$100	$75–$100 (includes jamb set and trim)	You can do this in 2 hours.
Installing a new fir wood paneled interior door	$600	$100	$400–$500 (includes jamb set and trim)	You can do this in 2 hours.
Installing a new standard 12′ garage door	$800 (Add $300 for automatic opener.)	$200	$400–$500	You can do this in 8 hours.
Installing a 12′ three-panel sliding glass door	$1800	$200	$1500	You can do this in 8 hours, with a helper.
Installing a 24′ x 60′ double-hung window	$250	$50	$200	You can do this in 4 hours.

UTILITIES

TASK	HIRING IT DONE		DOING IT YOURSELF	
	JOB COST	LABOR	MATERIAL	TIME INVESTED[5]
Installing 200-amp service run with interior breaker panel	$2800	$900	$1500	Unless you are a licensed electrician, you shouldn't try this.
Completely rewiring a 16′ x 24′ building	$2000	$1200	$600	You can do this in 24–40 hours.
Installing a 40-amp breaker sub-panel	$300	$80	$150	You can do this in 4 hours.
Installing a 120-volt outlet (15′ run)	$35	$25	$10	You can do this in 2 hours.
Installing a light and switch	$50	$30	$20	You can do this in 2 hours.
Installing rough and finish plumbing for a bathroom, complete with toilet, sink, and shower stall	$2000 total, with rough and finish plumbing costs about $1000 each	$600 for rough; $300 for finish	$150 for rough; $500 for finish	You can do the rough plumbing in 8 hours. The finish work will take 8 hours.

[5]Assumes an experienced worker

Restored outbuilding gallery

Here's a quick look at a few outstanding restored barns and outbuildings to inspire you. Not only do these pictures show that old barns can be successfully adapted to new uses, they also show that all the hard work and worry that goes along with a restoration really is worth the effort.

This small carriage barn once hosted "The Barn Gang" — a group of engineers and inventors whose accomplishments helped create the automobile industry. The electric starter, drum brakes, and octane gasoline were developed here. This barn was moved from its original location and restored in Moraine Historical Park in Dayton, Ohio.

This bank barn was restored to its original condition to become part of Carriage Hill Farm, a working 19th-century farm museum. Thousands of kids come to this museum each year to learn about our rural heritage.

This carriage barn was carefully restored, decorated, and painted to match the Victorian home it rests behind. It's now used as a garage — one of the few garages you'll find on the National Register of Historic Places.

This shed may be a common, nondescript building to most folks, but it held historical significance for the far-flung Studebaker family. In the 19th century, it served as a workshop on one of their pioneer homesteads. It was moved to its present location and restored as a working blacksmith shop.

These connected outbuildings were all part of a working farm. They once served as granaries, hay mows, and workshops. But when the farmhouse was made over into a bed-and-breakfast, these buildings were restored as crafts shops.

One of the earliest round barns, this huge stone structure is part of the historical Hancock Shaker Village in Massachusetts.

(continued on next page)

Restored outbuilding gallery — *continued*

This long barn was disassembled, moved to a new location, and reassembled as a meeting house for the WACO Historical Society and Aviation Learning Center.

This bank barn was made over into a club house at a country club. The downstairs holds a golf shop and a bar, while the equipment needed to maintain the greens and fairways is stored upstairs.

This round-roofed foundation barn was gutted, the post-and-beam frame was reinforced, and the structure was restored as a family dwelling.

Evaluating the structure

Once you've decided that a barn or an outbuilding is worth restoring or modifying, the next step is to carefully inspect the structure, evaluate it, and plan the project. This needn't be one of those chores for which you have to set aside a day and say to yourself, "Today I'm going to evaluate that barn." You own the structure, you probably use it on a regular basis, and you'll begin to notice things about it as you traipse in and out. Evaluation is usually a slow, thoughtful process. So is planning. As what you must do to restore the structure or adapt it to another purpose becomes apparent, the plan will grow of its own accord.

Type of structure

The first thing you should notice about the barn is the *type* of structure. What construction method are you dealing with? More than any other factor, this determines what you must do to make repairs. Each type of construction develops a different set of problems as it ages, and you have to know what to look for in your evaluation. *(See Figure 2-1.)*

The most common types of barn construction are:

◆ *Post-and-beam* — The barn is framed with large timbers and covered with plank siding. Post-and-beam barns usually rest on a stone foundation. *(See Figure 2-2.)*

◆ *Frame* — The barn is framed with dimension lumber and nailed together, much like a frame house. This is more common with smaller outbuildings and urban outbuildings. Frame buildings may sit on a variety of foundations — stone, concrete, or concrete block. *(See Figure 2-3.)*

◆ *Pole* — The barn is framed with chemically treated poles and dimensional lumber. There is no foundation; the poles are set directly in the ground. *(See Figure 2-4.)*

Log and masonry construction are less common, but just as important. *(See Figure 2-5.)* I'll refer to all of these construction methods throughout this book. Be aware that your barn probably incorporates several of these construction methods, especially if it has been around long enough to accumulate some repairs and additions. It may, in fact, be a patchwork quilt of construction methods that tells the history of rural building in your area. This will make it a pain to restore, but a joy to explore.

2-1. You can read this huge barn like a history book. The two end bays are post-and-beam structures put up in the 1850s. The huge middle bay that ties them together was framed in the 1920s. Some of the "bump-outs" use pole construction methods from more recent times.

2-2. Post-and-beam forebay barn

2-3. Framed carriage barn

2-4. Pole equipment shed

2-5. Stone grist mill

Post-and-beam construction

Traditionally, American barns were framed with posts and beams — large timbers, usually 8 inches square or larger, hewn from tall, straight trees. To put up a barn, settlers first laid a low stone foundation. On top of this, they laid *sill beams.* Then they assembled vertical *posts* and horizontal *tie beams* on the ground, making large frames called *bents.* An extremely long bent might have one or more posts between the outside posts for additional support.

The carpenters raised the bents in place on the sill, then tied them together with *girts* between the posts. They reinforced the assembled timber frame with diagonal *braces* between the horizontal and vertical members.

The top girts and the top tie beams became the *plate* on which the carpenters installed the roof. They raised the *rafters* to the proper slope and tied them together at the peak with a *ridgepole.* It was also common to tie pairs of rafters together with a *collar.* If the rafters were so long that they needed support in the middle of their spans, horizontal *purlins* were installed to help support the weight. The purlins themselves were supported, usually by several extended posts. Posts that supported purlins were called queen posts. A post that extended all the way to the peak of the rafter to support the ridgepole was a king post.

All of these timbers were assembled with *mortise-and-tenon joints.* The surface of one timber was drilled and chiseled to make a rectangular slot, or *mortise.* The end of the adjoining timber was shaped to make a tongue, or *tenon,* to fit the mortise. The carpenters drilled holes through the mortises and tenons, then locked them together with *pegs.* Nails were rarely used in post-and-beam construction.

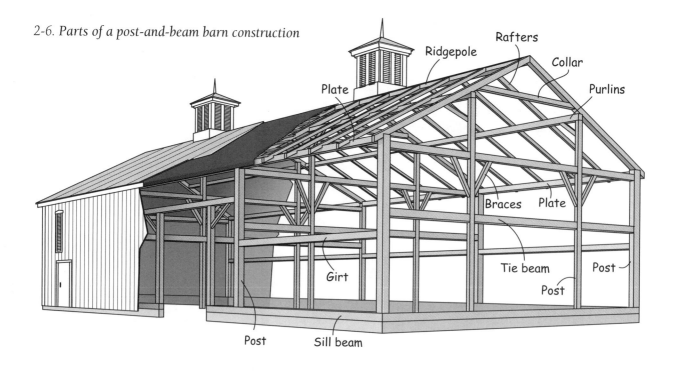

2-6. Parts of a post-and-beam barn construction

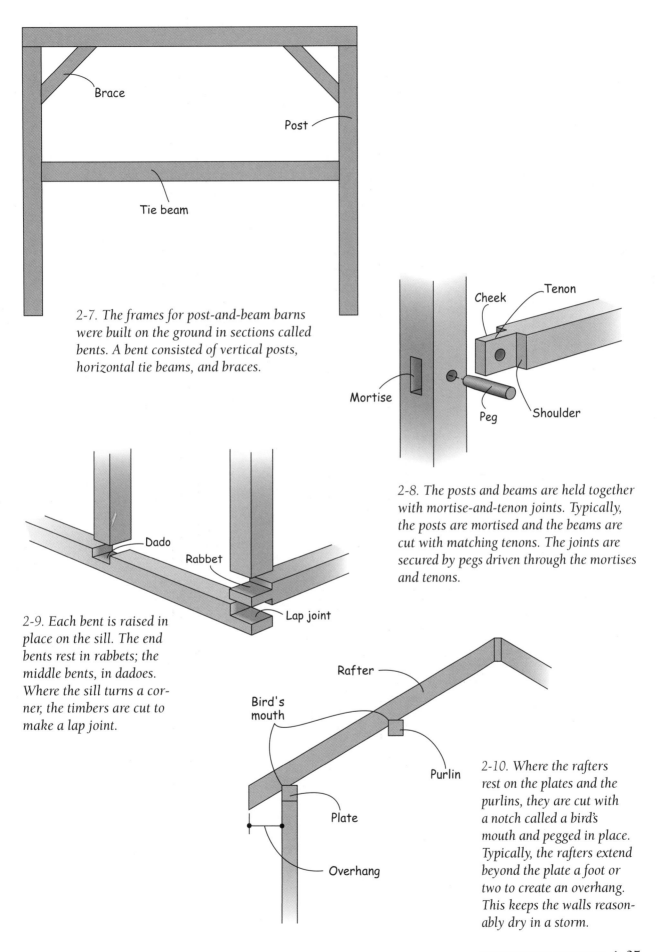

Brace

Post

Tie beam

2-7. *The frames for post-and-beam barns were built on the ground in sections called bents. A bent consisted of vertical posts, horizontal tie beams, and braces.*

Cheek

Tenon

Mortise

Peg

Shoulder

2-8. *The posts and beams are held together with mortise-and-tenon joints. Typically, the posts are mortised and the beams are cut with matching tenons. The joints are secured by pegs driven through the mortises and tenons.*

Dado

Rabbet

Lap joint

2-9. *Each bent is raised in place on the sill. The end bents rest in rabbets; the middle bents, in dadoes. Where the sill turns a corner, the timbers are cut to make a lap joint.*

Rafter

Bird's mouth

Purlin

Plate

Overhang

2-10. *Where the rafters rest on the plates and the purlins, they are cut with a notch called a bird's mouth and pegged in place. Typically, the rafters extend beyond the plate a foot or two to create an overhang. This keeps the walls reasonably dry in a storm.*

Frame construction

As the tall trees needed for massive posts and beams became scarce, rural carpenters began to use a new building technique that used smaller-dimension lumber. Frame construction used "two-by" lumber — boards cut 2 inches thick and ripped to various widths. Instead of making bents, carpenters framed walls. They made the wall frames on the ground, nailing the parts together. Vertical *studs* joined a *top plate* and a *sole plate* to make a rectangular frame. To make an opening for a door or window, carpenters cut the studs short and spanned the opening with a thick *header* that would support the weight of the structure above it. *Jack studs* supported the header. These were nailed to *king studs* on either side of the opening. The shortened studs were referred to as *cripple studs.*

When the wall frames were complete, carpenters raised them in place on the foundation and tied them together at the corners. Usually, they nailed a second top plate to the first to create a double plate at the top. The second top plate spanned joints where the members of the first top plate butted against one another. This helped tie the wall frame together.

2-11. Parts of a frame barn construction

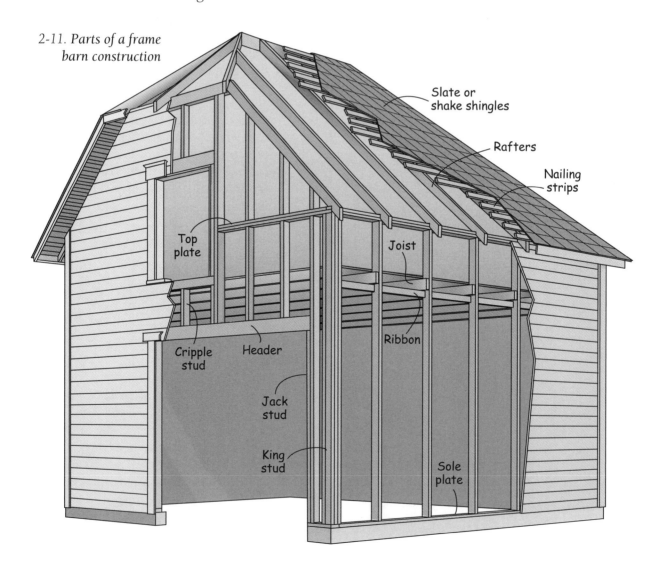

Slate or shake shingles

Rafters

Nailing strips

Top plate

Joist

Cripple stud

Header

Ribbon

Jack stud

King stud

Sole plate

The carpenters installed the roof much as they would install a roof on a post-and-beam barn. They leaned *rafters* against a ridgepole and tied the rafters together with *nailing strips*. Later, the nailing strips would support the roofing materials.

To install a loft, the carpenters cut dadoes in the studs and "set in" horizontal *ribbons*. They stretched joists across the ribbons, nailing the joists to the studs.

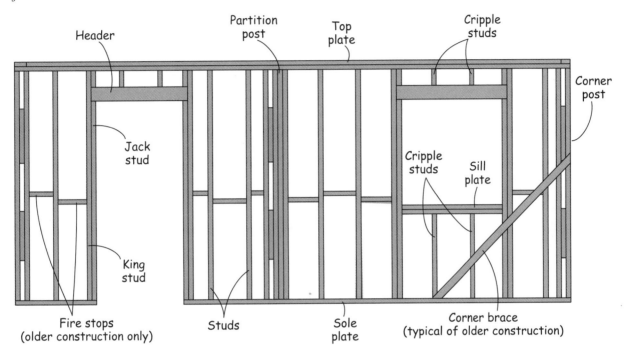

2-12. *A typical wall frame with openings for windows and doors. The headers are typically 2x8s turned on edge, although wider openings may require larger boards. The rest of the frame members are 2x4s.*

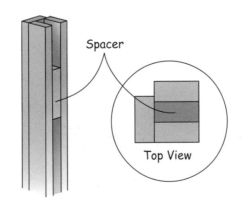

2-13. *Where two frames join to turn a corner, carpenters nail the studs edge to face. They also install a third stud, as shown, to create an inside corner. This is essential if you plan to cover the inside walls. It's not absolutely necessary on outbuildings where the interior is left raw, but it does strengthen the corner.*

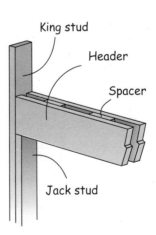

2-14. *To create a header, nail two wide boards together with spacers between them. The completed header must be as wide as the studs.*

Pole construction

The newest wrinkle in barn construction is the pole barn. Round or square poles are erected vertically, much like the posts in a post-and-beam structure. The poles sometimes rest on large stones or concrete piers, or they may be set deep in the ground so the bottom ends are well below the frost line. Manufacturers of building materials treat barn poles with chemicals to prevent rot when in direct contact with the ground. The poles are tied together with horizontal *girts*. At the top and bottom of the structure, carpenters typically attach both inner and outer girts to reinforce the barn. The girts at the very bottom of a wall are sometimes referred to as *splashboards*.

Doors and windows are framed much as they are in a frame structure. Carpenters nail inner and outer *headers* to the poles, then attach *casing boards* to the poles and headers. The edges of the casing boards are even with the surfaces of the inner and outer girts.

The roofs on pole barns are normally trussed. The carpenters assemble triangular frames or *trusses* on the ground from dimensional lumber. Each frame consists of two *rafters*, a horizontal *chord*, and some interior braces or *web members*. The parts of the truss are fastened together with nails and *gussets*. When the trusses are complete, they're lifted in place on the top girts and tied together with *roof sheathing*. Later, the sheathing is covered with roofing material and the walls are covered with siding. Older pole barns are usually covered with wooden board-and-batten siding, newer ones are clad with metal siding.

2-15. Parts of a pole barn construction

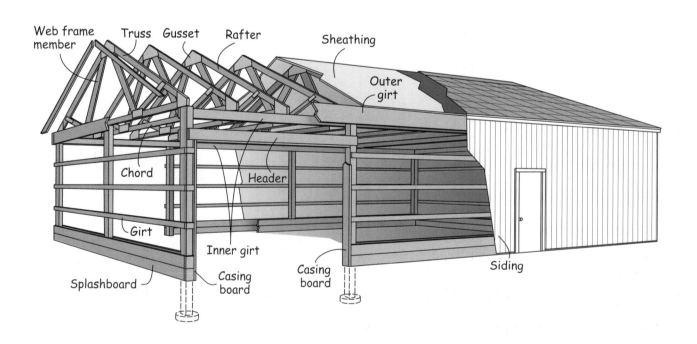

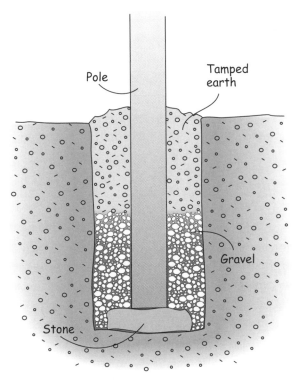

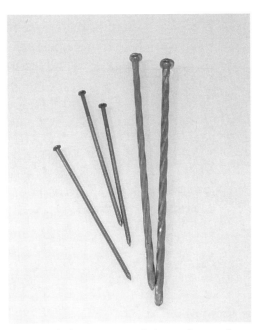

2-16. *Barn poles that are set directly in the ground should rest on a large, flat stone or concrete block to keep them from settling. There should be gravel around the bottom of the pole to help drain the water away from the wood.*

2-17. *Pole barns are nailed together with the mother of all nails, pole nails. While these make the structure strong, they are also difficult to remove. This can add time and effort to a restoration or repair.*

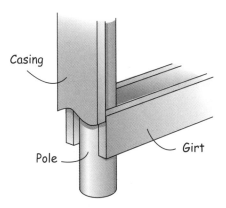

2-19. *Casing boards finish off an opening and provide a solid surface in which to anchor door or window casings.*

2-18. *The roof trusses must be securely attached to the posts — not just toenailed into the girts. Otherwise, the roof can blow away in a high wind. Make it a point to check this when you evaluate the structure.*

Log construction

Log outbuildings are exceedingly rare. For the most part, these were temporary barns and sheds that were used only until the settlers could afford better. Often, they were pulled down and the timbers were used to help frame a post-and-beam barn. But a few of them are left, sometimes hidden under siding that was applied at a later date. If you are one of the anointed who has inherited one of these beautiful old structures, you have a valuable piece of history on your hands.

2-20. The dovetail-shaped joints on the ends of the logs prevent the logs from slipping out of place.

Log structures were built by laying up logs horizontally to make a rectangular "crib." To make the structure rigid, the ends of the logs were notched to lock together. The pioneers used several different types of notch joints, but the most common was a *dovetail notch*. *(See Figure 2-20.)* Once the logs were in place, the builders cut a door opening and framed it in with *casing boards*. The roof was installed in much the same manner as the roof on a post-and-beam barn — the builders rested *rafters* on the top logs. The rafters were joined at the peak with a *ridgepole*.

The spaces between the logs were filled with *chinking*. The builders stuffed short, split logs — firewood — into the opening, then plastered this over with clay. Occasionally, the spaces were stuffed with rocks or moss. Since this was in many ways a "make-do" structure, the builders used whatever materials they had on hand.

2-21. Parts of a log barn construction

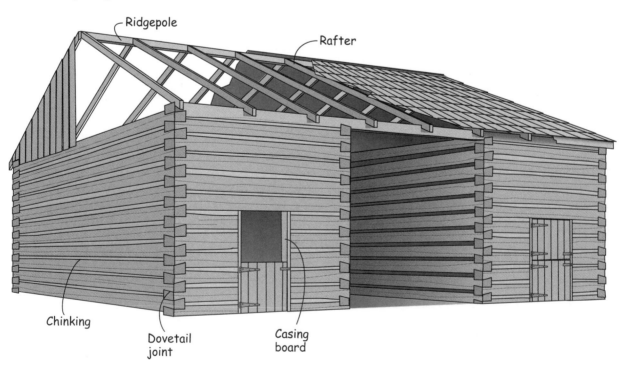

Masonry construction

In areas where masonry materials were cheap and abundant, barns were built from stone, brick, block, or tile. In the northeastern United States, for example, barns made of "fieldstone" are common. When the glaciers retreated from this area after the last ice age, they left an abundance of small and medium-sized rocks lying around the landscape. Thousands of years later, the settlers began gathering these rocks and building structures with them. To give you a more specific example, there used to be a huge tile factory in southeastern Ohio. Consequently, you can find many barns and outbuildings in the neighborhood built from drainage and sewer tiles.

A masonry structure is made by laying up stones, bricks, or other materials with mortar. The masons built the openings into the walls as they laid them up. To finish off an opening, they typically laid a large wooden *lintel beam* across the top, then continued to lay masonry on top of the lintel. Occasionally, the masons created an arched opening, laying up the materials around a wooden form and then removing the form when the opening was complete.

Once the walls were up, carpenters built a roof in the same way they would build a roof on a post-and-beam barn — *rafters* leaning against a *ridgepole*. Sometimes, the gables were closed in with more masonry; other times, they were framed and sided with lumber and boards.

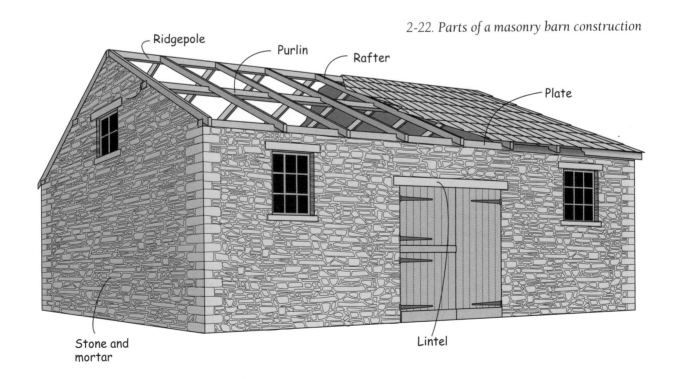

2-22. *Parts of a masonry barn construction*

Ridgepole

Purlin

Rafter

Plate

Stone and mortar

Lintel

A barn dictionary

Barn construction, like any craft or technology, has a language all its own. When restoring a barn, it helps to know the terms. The good folks at the building supply store will be forever grateful if you show up knowing what to ask for. Early in my career, I had a conversation with the owner of a feed and farm supply store that went something like this:

"I need to replace one of those vent-things on a barn."

"Vent-things?"

"Yeah — they let the air out of the barn."

"Do they let the air out of the gable or the ridge?"

"Um ... the air comes out of the very top."

"You mean the ridge."

"I think so."

"You need a ridge ventilator. Now, would you like a standard ridge ventilator, a forced-air ridge ventilator, or a cupola ridge ventilator?"

At this point, I surrendered and asked to see the catalog so that I could just point to what I wanted.

To help improve your communication skills and avoid possible misunderstandings, here's a quick lesson in barn terminology.

Surfaces

Most barns have a roof and four walls. The lowest edges of the roof are the *eaves;* the peak where the roof surfaces join is the *ridge*. The walls under the eaves are the *sides* of the barn. The portion of the wall that fills the space between the roof surfaces just under the ridge is the *gable,* and the portion of the wall under the gable is the *end*. When referring to the entire wall, carpenters sometimes call it a *gable end*.

Interior

The large door on the side or the end of the barn leads to the *runway* — an open aisle that usually goes down the center of the barn. On either side of the runway, there may be *stalls* for livestock or *granaries* for storing grain. Above the stalls, there is a *loft* for storing hay.

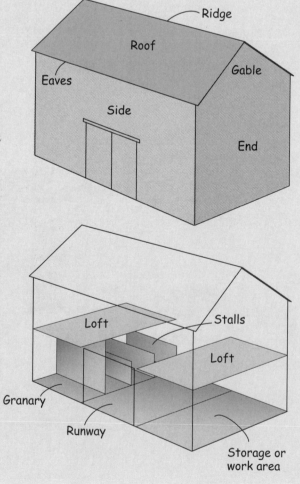

Roofs

The most distinctive thing about a barn is the shape of its roof.

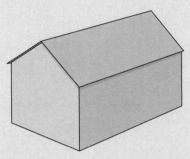

A common *gable roof* has two sloped surfaces that join at the ridge.

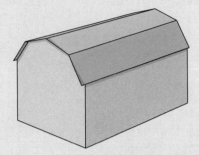

A *gambrel roof* has four surfaces. The top two are at a shallower pitch than the bottom two. This increases the storage space in the barn loft. This roof design got its name because of its resemblance to the shape of an old butcher's hook, which is called a *gambrel* in French.

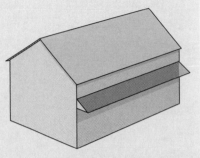

A *pent roof* is an awning on the outside of the barn to provide additional shelter.

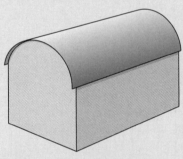

A *round roof* (also called a *rainbow roof*) is arched so that it has no ridge. These first appeared on barns early in the 20th century.

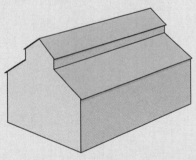

A *monitor roof* or *top-hat roof* has a raised section in the center. Often, this section has windows along the short walls between the roof surfaces to help light and ventilate the barn. These roofs are not often used on large barns, but they are common on smaller outbuildings, horse barns, and meadow barns.

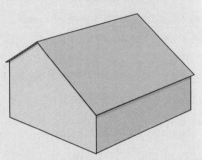

A *saltbox roof* is a gable roof with one surface much longer than the other. Sometimes, a barn assumes a saltbox profile when a shed is added to one side.

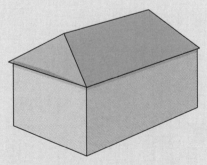

A *hip roof* has a sloping roof surface on all four sides. These roofs are not widely used on rural barns because the hip design decreases the loft area, but they are common on urban carriage barns.

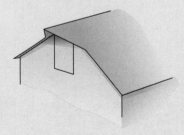

A *hay hood* or *hay bonnet* is a gable extension near the roof ridge. This helps shelter a gable opening. The hay hood is often fitted with a block and tackle to help load hay into a loft.

(continued on next page)

A barn dictionary — *continued*

Walls

The walls of barns are usually covered with vertical boards called *siding*. The gaps between the boards are covered with small strips of wood, or *battens*, to help weather-proof the structure. This arrangement encourages the rainwater to run off and offers few horizontal surfaces where water can collect. Occasionally, boards are lapped horizontally. This is *weatherboarding*. Carriage barns and garages are often sided with horizontal *clapboards* to match the homes they belong to. These narrow boards are beveled from top edge to the bottom edge, and may fit together with molded tongue-and-groove or tongue-and-rabbet joints.

Board-and-batten siding

Clapboard siding

Weatherboard siding

Openings

Windows and openings in a barn serve two purposes. They provide light to work in the barn, and ventilation to dry the hay and the grain.

Cupola ventilators are small enclosures that protrude from the ridge of the roof. The walls of the enclosure have openings or louvers to let the air escape. Often, these cupolas are decorative as well as functional.

Ridge ventilators also protrude from the ridge, but they are made of metal. Usually, they have a round shaft.

A *gable ventilator* is an opening at the top of the gable, near the ridge, to allow air to escape. Often, this opening is protected by a bargeboard that stands out from the wall a foot or so. The bargeboard prevents the rain from entering the ventilator opening.

Dormers are enclosures set in the slope of the roof to allow light in and air out. Often, the end of the dormer holds a window. These are more common in carriage barns and garages than rural barns.

Owl holes are small cutouts in the gable ends, near the ridge. They often have a decorative shape. Not only do they allow light in and air out, they also provide passage for owls and other raptors. This helps control the mice populations in barns that have granaries.

Louver ventilators sometimes look like shuttered windows set in the wall of a barn. There is no glass in the frames, however, just wooden louvers sloped to prevent the weather from entering the barn without interfering with the air flow.

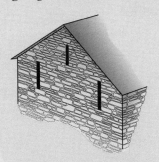

Slit ventilators are long vertical slots in barn walls, like a very tall, very narrow window. They are most common in German barns, particularly ones made from stone or brick.

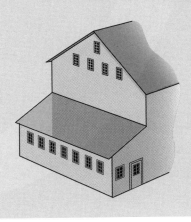

Barns may also have *windows* to provide light and ventilation. They tend to have less of them than other structures, and these are often concentrated in just one part of a barn where the farmer has set up a workroom. The exception to this rule is dairy barns — these often have rows of windows along the ground floor where the cows are milked.

(continued on next page)

A barn dictionary — *continued*

Access

Barns typically have doors of several different sizes. The large *barn door* in the side or end leads to the runway. Originally, these doors were made big to let a farmer drive in wagons loaded with loose hay and pitch the hay up to the loft. Nowadays, the large doors are needed to accommodate large pieces of farm equipment. In addition to the barn door, barns may have one or more smaller door to let humans and animals in and out. Occasionally, these smaller doors are cut in two horizontally to make a *Dutch door.* The Dutch door allows the farmer to open the top portion for light and ventilation, but leave the bottom portion closed to keep animals from wandering in or out. A *loft door* covers an opening in the gable that lets a farmer load and unload hay from the end of a barn.

Barn door

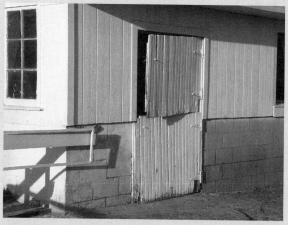

Dutch door

Loft door

End use

Once you understand how your barn or outbuilding is constructed, you must give some thought to how the building will be used after you've renovated it. There are five common uses for outbuildings, and each of them requires specific improvements or changes to the basic structure.

Humans, full time

If people will be living and working in the structure all day, you will want all of the amenities that a home requires.

◆ *Electricity* — The building must have adequate power to run all the appliances you'd find in a home. Most modern homes have 200-amp service with both 110-volt and 220-volt circuits.

◆ *Plumbing* — You must have an adequate supply of water, either from a well or a municipal system, and a drain waste vent system tied to a municipal sewer or a septic tank and field. You'll also need a water heater. If you choose a natural or propane gas water heater, you must also install gas lines and tie them either to a municipal gas utility or a propane tank.

◆ *Heating and cooling* — You'll need an adequate heating ventilation and air-conditioning (HVAC) system, and the utilities to run it — electricity, gas, or oil.

In all probability, you'll also have to make many changes to the structure of the outbuilding. For example, if you're adapting a barn to live in, you'll need to add walls, ceilings, floors, and stairs. The structure will most likely have to be strengthened to support these new additions. You will have to add windows — barns don't usually have as many windows as people commonly have in their homes. And you'll have to insulate the outside walls and install airtight doors so the heating and cooling system will work efficiently.

2-23. This round-roofed barn was adapted to become a home. The original structure was built in the 1930s as a dairy barn. The owners gutted it and reinforced the frame. They added an entirely new interior structure to divide the space into rooms.

Humans, part time

If you're restoring the outbuilding to make a workshop, a studio, or a similar structure, then people will be using it only part of the time. What utilities you add and how you adapt the structure will depend on the activities that will take place inside the restored building, your geographical location, and the times of the year during which you want to use the building. A woodworking shop in the northern United States that you expect to use only during the warmer months may need only electrical utilities. If it's just a hobby workshop, you may be able to run an electrical power line from your home to a sub-box in the outbuilding. If you're using the building for an activity that requires water, such as photography or pottery, you may need some simple plumbing — just a faucet and a sink. On the other hand, if the outbuilding will house a business in which several people will work during the day, local codes and common sense will require you to install a separate electrical utility, toilets, heating, and air-conditioning.

Here again, expect to make some changes to the structure by adding interior walls, floors, ceilings, doors, windows, and insulation. Occasionally, you may have to remove or replace structural members. Typically, when I restore an outbuilding to make a workshop, I remove the interior support posts to create an open, unobstructed work area. To do this, I have to replace the old wooden beams with steel I-beams, trusses, or "MultiLam" plywood beams that can span a wide area without intermediate supporting posts.

2-24. This is the interior of a carriage barn that I restored to make a professional woodshop. It has its own electrical service and a heating and air conditioning system. I kept the barn doors so the owner could move large assemblies in and out of the shop, but I insulated the doors and made them airtight.

Animals

If you want to adapt the structure to house animals, bear in mind that each type of animal has specific needs. I've prepared some guidelines to help you plan a shelter for common farm animals, but bear in mind these are far from complete. I haven't the space to provide all the information you need to keep an animal comfortable and healthy — you must do some additional research. Start by asking someone who keeps the same animals you plan to keep and find out firsthand what's involved. Follow this conversation up with a visit to your local library, veternarian, or agricultural extension office to find more information and sample plans of animal shelters.

Chickens — A chicken coop must be easy to clean, must provide good drainage, and must protect the chickens from wind, sun, rodents and other predatory animals. You will need adequate space for the flock — it must be well ventilated, yet free of drafts. Also supply natural and artificial light and enough space for sanitary feed and water stations.

You could keep 30 to 50 chickens in a room that measures 8 feet by 8 feet by 8 feet. The coop must be ventilated in hot weather and heated in cold weather — chickens thrive at 45 to 80 degrees Fahrenheit. These fowl also need an outside area with easy access to the coop. To protect the chickens from predators, both walking and flying, this area should be fenced in and fenced over.

A nest for an egg-laying chicken requires a 12-inch by 12-inch by 12-inch space. This makes for easy planning. One common plan for nesting areas is to install easily removable racks or shelves about 12 inches apart along one or more walls. Build nests just above the shelves and cover the bottom with hardware cloth so the feces pass through. As the chickens dirty the shelves, you can remove, clean, and replace them without disturbing the laying hens.

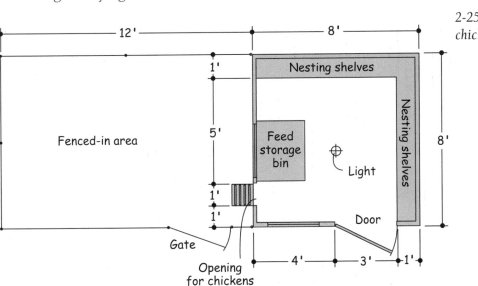

2-25. Layout of a small chicken coop

In some parts of the world, at certain times of the year, chickens have to be "tricked" into laying eggs. To do this, you must use artificial light to extend the days in the winter months. A lighting source also makes it easier to clean the coop and care for the chickens. A 75-watt light bulb 7 feet off the ground will provide light for 200 square feet of chicken space. You'll also want electrical outlets and heaters for the cold months.

Depending on the size of your chicken ranch, you may require separate spaces for a brooder, incubator, and slaughter room, all of which have different space and utility requirements. If you have set your poultry ambitions this high, I have to refer you to a poultry professional. Get some advice from someone who's been there to help you plan your outbuilding.

Horses — A horse barn should provide a safe, comfortable, and healthy equine home. In the summer, you'll need plenty of sunlight and fresh air. In the winter, you should be able to make the barn reasonably airtight to protect the horse from drafts. In extreme northern latitudes, you may need to provide supplemental heat.

The stalls in a horse barn must allow the horse to move freely. A 12-foot by 12-foot space provides adequate room for a normal size horse. Larger breeds, particularly draft horses, may require more space. The stall walls can be made of wood, stone, brick, or cement block but must be able to withstand kicking, rubbing, chewing, and the chemical effects of urine and manure. The top section of a stall wall (from about 5 feet above the floor) should be either chain mesh or wooden bars with no more than 2 inches between each bar. This allows for ventilation and lets the horse look out. The ideal stall floor is about 1 foot of well-drained topsoil.

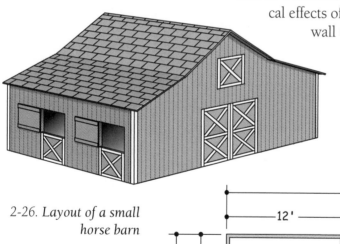

2-26. Layout of a small horse barn

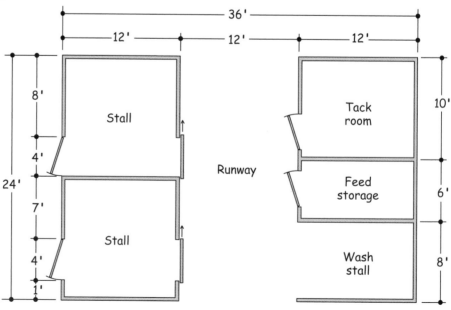

In addition to the horses, tack must also be kept in the barn. A small horse barn (two to four stalls) normally requires 100 to 200 square feet of storage space for tack. A tack room should be well organized and easily accessible from the grooming and saddling area. Since you will be storing leather goods in this room, it should be well protected against rodents. For this reason, you'll want to install a wooden or concrete floor. To keep the leather in optimum condition, you should maintain about 60 percent relative humidity. If you store medicine in the room, you may want to insulate the room and install a small space heater to keep the medicine from freezing.

The barn must also store at least a few weeks' supply of hay and feed for the horses. Like the tack room, the feed room must also be free of rodents, with a hard floor and gnaw-proof bins. Metal garbage cans make excellent grain bins for small horse barns.

Cows — A small cattle barn should provide a place to protect the cows in bad weather. It should also have a calf stall to protect newborns and, if you keep dairy cattle, a milking area. This structure can be very simple — the cows won't care since they spend most of their time outside. A simple 16-foot by 20-foot shed makes an adequate barn for four to six cows. The interior can be partitioned to make a calf stall and a milking area. The floor of the barn, with the exception of the milking area, should be well-drained top-soil . The milking area should rest on a concrete pad to make it easy to clean. A milking area for a single cow is normally about 3 feet wide and 8 feet long.

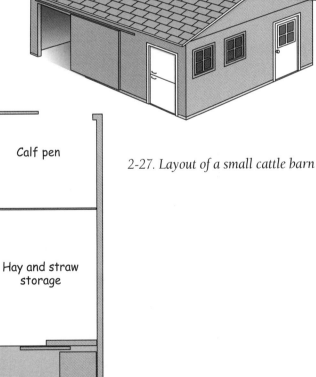

2-27. *Layout of a small cattle barn*

Calf pen

To lot or pasture

Shed

Manger

Hay and straw storage

20'

Drain

Milking station

Manger

To lot or pasture

Granary and milking equipment

16'

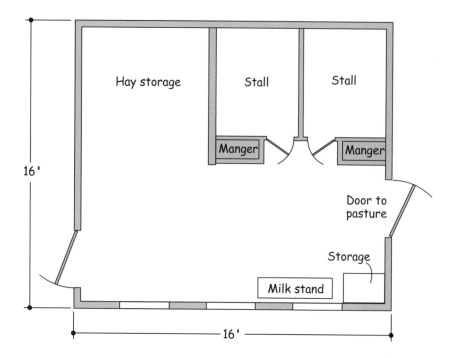

2-28. Layout of a goat house

Goats — Goats are probably the easiest of all farm animals to house because they will live anywhere. I once kept a lovable old goat happy and comfortable in a large doghouse. In fact, I suspect he thought he was a dog — he would come when I whistled. However you make your goat house, it should be dry and free of drafts, as goats are susceptible to pneumonia. Ideally, it should be well ventilated, should be well lit, and should have windows with a southern exposure. You may want to partition the goat house to make one or more box stalls for nursing nannies and their kids. If you milk the goats, you'll need a milk stand and space to keep it. The floor should be well-drained topsoil. You don't necessarily have to pour a concrete pad for the milk stand so long as the stand itself has a hard, smooth platform that can be cleaned easily.

Pigs — I hesitate to devote much space to pigs for the simple reason that pigs are best kept in a movable structure without a foundation. I doubt many people will refer to this book in order to renovate an outbuilding to keep pigs. Most pig houses are low to the ground — no more than 5 feet tall — and mounted on skids. Pigs make a terrible mess. The sharp hooves and the massive weight tear up the ground beneath the pig house, mixing the urine and feces deep into the earth. Rather than clean the floor, many pig farmers find it more expedient to hitch the pig house to a tractor and drag it over to a fresh, clean spot on the pig lot.

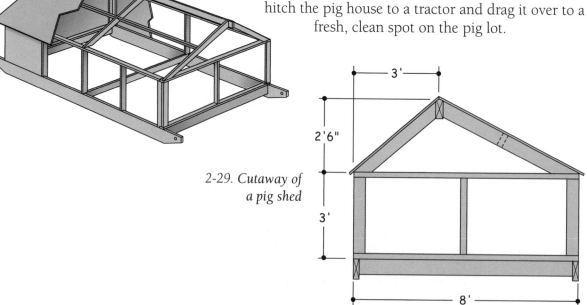

2-29. Cutaway of a pig shed

Automobiles and machinery

If you want to use an outbuilding as a garage, it must have adequate access — in other words, a garage door. The same applies if you want to keep tractors, wagons, and other mobile machinery. Next, you need a strong floor to support the weight of the vehicles or the machinery. Wooden floors must be properly braced. If you're restoring an old bank barn with the intention of keeping a large tractor in the upper runway, you may have to install floor jacks or supporting posts under the floor joists. If the barn rests on a concrete pad, or you intend to pour a pad beneath the barn, the pad must be thick enough to support the machinery. Normally, garage floors are at least 6 inches thick.

Also give some thought to moisture control. As moisture condenses on a cold metal surface, it rusts your machinery. To prevent this, you must keep the air temperature inside the building a few degrees above the outside air temperature. There are small electric heaters, sometimes called pump house heaters, that can raise the temperature a few degrees inside a small barn.

Do you want to maintain the automobiles or the machinery that you will be keeping in the restored outbuilding? If so, you'll want storage space for the tools you need to do the maintenance. Often, folks incorporate a small machine shop inside the building where they keep their equipment. I once helped a gentleman rebuild a barn for his collection of historic automobiles. When we poured the concrete floor, he installed an old-time grease pit — a sunken area in the floor — so he could more easily work under the autos.

2-30. This carriage barn was restored as a garage to hold several automobiles. If you make a similar conversion, measure the interior of the barn carefully. Nineteenth-century carriages were somewhat shorter than modern cars, so you may have to enlarge the barn to hold your auto. I've seen people add small sheds to the side of a carriage barn to provide enough space to park a car.

Storage

Many outbuildings are used to store stuff. *Stuff* is an all-inclusive word that incorporates an enormous number of inanimate objects. And some of this stuff has specific storage needs. Do you need to keep the stuff from freezing? Does the structure need to be ventilated so the stuff won't mold or mildew? Is the stuff sensitive to light or humidity?

One of the most overlooked needs in storage buildings is organization. How do you group the stuff so you can find it again? How should it be stored so you can easily retrieve it without moving a whole lot of other stuff? Often, a simple system of built-in shelves or racks will save you some storage headaches down the road.

2-31. This small barn was originally built to hold lawn and garden equipment, and then I adapted it to hold lumber. The racks separate the board by wood species. I tore out the floor and replaced it with strips of wood spaced ½ inch apart. I also installed ventilation grates in the gable. This keeps the air moving through barn, maintaining the wood at equilibrium with the prevailing relative humidity.

Building codes, permits, and inspections

Okay, we covered all the fun stuff. Now we have to deal with your local bureaucracy. There comes a time in every major restoration when you have to consider the local building codes and inspection procedures that were put in place to protect you and your neighbors, and this is it. Your local building codes sound about as exciting as a tax form, I know. But every cloud has its silver lining, and in this case the lining is invaluable. I have yet to meet a county engineer or building inspector who didn't teach me something. By and large, they are a helpful bunch, full of good advice, tips, resources, and contacts. There's a good reason for this helpfulness — their job depends on you completing your restoration safely and successfully.

Building codes

Take a trip to your nearest county or city building and ask to talk to the resident engineer about building codes. Tell him or her that you are planning a restoration and ask what codes apply to your type of building. In most cases, you will be pleasantly surprised that the codes that apply to a restoration are a lot less prohibitive than those that apply to a new structure. Materials, techniques, even building locations are often considered to be "grandfathered" in old structures, and as long as you don't change them substantially, the codes for new construction won't apply. In some cases a "50 percent rule" may apply. This can be interpreted differently by municipal agencies. One definition is that if you are spending more than 50 percent of the value of the building as it stands before you start (land value not included), then you have to bring the entire structure up to code.

However, don't take this as gospel — codes differ quite a bit from one location to another, and you should check them out before you do any renovating. The time you spend with the local inspector or building standards department is an investment to hedge against later frustration.

I know a gentleman from Atlanta whose storage barn was removed by a tornado. He rebuilt it just the way it was, on the same foundation, and the authorities made him tear it down. It wasn't to code!

Also, while you're at the engineer's offices, find out what rules regulate trash disposal. A renovation not only generates tons of trash, it also generates an amazing variety of trash — wood, stone, glass, shingles, concrete, asbestos, horseshoes, hubcaps, an occasional skeleton of a long-dead animal. Find out how each type of trash should be dealt with. And look for the silver linings — I have actually made money salvaging aluminum siding from a structure. This helped offset the cost of the building materials I had to purchase.

Building permits

Depending on your structure, where it's located, and what work you plan to do in the course of the restoration, you may need a building permit. The rule of thumb is that you need a permit if you're altering or making major repairs to the structure of a building and that building is within the limits of a municipality. Rural structures are often exempt from the laws that require permits for the simple reason that those laws don't apply outside the boundaries of the municipality that made them. However, this pattern is changing rapidly: Many counties now require building permits for specific types of construction, repairs, and improvements. It's best to check.

2-32. You must display the building permit at the work site. This helps inspectors to find you and track your progress.

Applying for a building permit involves filling out a form and paying a small fee to have it processed. (I've included a sample form on the next page so you can get some idea of what's involved.) In some cases, you may have to supply the city or county engineer with a set of architectural plans to show him or her exactly what you intend. This is often a worthwhile exercise because it forces you to think through the job that you're about to do. If, while you're cogitating over the permit application, you come across a step in the restoration for which you have no clue about how to proceed, ask the engineer for advice. Show him or her this book and explain that the author of it doesn't have all the answers. The engineer will be tickled to help you out.

Once your application has been processed, you'll be issued a building permit. This must be displayed at the site while you are renovating the structure. The engineer should also give you a schedule of inspections, with instructions on whom to call when you arrive at the milestone in your work when you will require an inspector.

Preliminary inspection

To properly plan a restoration job, you should begin by thoroughly inspecting the structure. To help you out, here's a list of general problems that you should look out for. You may have your own list of specific features that are important to you.

Sample building permit application

CITY OF TIPP CITY   OFFICE OF BUILDING REGULATION   **BUILDING PERMIT NUMBER**

_____

By _____   Zoning Permit No. _____

Street and No. Location _____   Zoning District _____

Lot No. _____   Subdivision _____

NOTE: SEPARATE APPLICATION AND FEE REQUIRED FOR INDIVIDUAL BUILDINGS ON SAME PROPERTY.

TYPE OF IMPROVEMENT:

☐ New Building   ☐ Additions — Enter number of square ft. _____   ☐ Alteration — Enter number of added or deducted square ft. _____

☐ Repair, replacement   ☐ Moving   ☐ Describe briefly proposed work _____

RESIDENTIAL No. of Stories:

☐ One Family   ☐ Two Family   ☐ Three Family   ☐ Car Garage (☐ Attached)   ☐ Car Port (☐ Detached)

☐ Other_____

ADDITIONAL INFORMATION

☐ Basement   ☐ Concrete Floor   ☐ Crawl Space   Foundation: ☐ Concrete Block   ☐ Concrete   ☐ Fireplace

Roof Type: ☐ Asph. Shingle   ☐ Built up   ☐ Wood Shingle

COST (Omit Cents)

Estimated cost of Improvement for which this application is being made: $ _____

Square footage of residence and attached garage_____

Square footage of uninhabitable basement _____

Basic Fee .... _____

Fireplace Fee .... _____

Basement Fee .... _____

Accessory Building Fee .... _____

Dwelling and Attached Garage Fee .... _____

TOTAL _____

COMPLETE ALL ITEMS FOR NEW BUILDINGS AND ADDITIONS ONLY

PRINCIPAL TYPE OF FRAME   TYPE OF HEATING FUEL   TYPE OF WATER SUPPLY

☐ Masonry (wall bearing)   ☐ Gas N.   ☐ L.P.   ☐ Warm Air   ☐ Public

☐ Structural Steel   ☐ Oil   ☐ H. Water   ☐ Private (Well, cistern)

☐ Wood Frame   ☐ Brick Veneer   ☐ Coal   ☐ Radiant   FOR RESIDENTIAL BUILDINGS ONLY

☐ Reinforced Concrete   ☐ Electricity   ☐   Number of bedrooms _____

☐ Other_____   ☐ Other   ☐   Number of bathrooms _____

Is there Central Air Conditioning In   TYPE OF SEWAGE DISPOSAL

this building? ☐ Yes ☐ No   Size_____   ☐ Public Sewer

☐ Private System (septic tank, etc.)

IDENTIFICATION	NAME	STREET ADDRESS	CITY	STATE	ZIP	PHONE NO.
OWNER						
CONTRACTOR						
PLUMBER						
ELECTR. CONTR.						
HEATING CONTR.						

In consideration of the issuance of this permit, the owner and his agent or contractor do hereby covenant and agree to comply with all laws of the State of Ohio and the Building Code and Zoning Ordinance of Tipp City, Ohio, and to install the proposed building and/or work, or make the proposed change or alteration or do the work described above, in accordance with the plans and specifications as approved by the Building Inspector, and certify that the information and statements given on this application and the accompanying drawings and specifications are true and correct to the best of their knowledge.

Application by_____ Address _____ Phone _____

OWNER'S OR AGENT'S SIGNATURE

State of Ohio, Miami, ss:

Sworn to before me and subscribed in my presence this _____ day of _____ 19 _____

_____

NOTARY PUBLIC

Foundation — Walk around the structure and follow the foundation with your eyes.

◆ Is it cracked anywhere? If so, how badly?
◆ Has the foundation shifted toward or away from the building? Are the framing members sitting on the foundation?
◆ If your foundation is stone, do the stones look to be in place? Have they shifted?
◆ Have the roots of trees growing near the structure affected the foundation?
◆ Is the foundation strong enough to support the structure? In many cases, buildings were erected without sufficient foundations, or rooms were added to create a larger structure whose weight is too much for the original foundation.
◆ Is the roofline sagging or humped? Looking along the roof may help you identify a problem that begins with the foundation.

Floor — Get down on your hands and knees and sight along the surface.

◆ Is the floor sagging, bowed, or slanted? Is the problem the joists or a settled foundation?
◆ Do the joists show signs of dry rot or termite damage?
◆ Are the floorboards broken, missing, or rotted?
◆ Is the crawl space beneath the floor properly ventilated?
◆ If the floor is concrete, is the slab cracked or crumbling?
◆ Is the floor strong enough to hold any equipment you want to store?

Frame or structure — Walk around both the outside and inside of the building.

◆ Is the structure leaning? (This is a very common problem with barns.)
◆ Do the frame members show signs of dry rot or termite damage?
◆ Are any of the frame members broken?

EXTRA HELP

To do a good inspection, you'll need a simple tool kit. I keep mine in an old backpack and use it whenever I need to check out a building.

◆ *Old straight-edge screwdriver* — Makes a great scraper, pry bar, door opener, wood rot gouging tool, mortar strength inspector, etc.

◆ *Sharp pocketknife or utility knife* — Excellent for inspecting rot and scraping away weathered or painted wood to determine its species and grain. Also useful for cutting through tangles of vine.

◆ *25' tape measure* — 25' seems long enough for me to measure anything that I need to measure.

◆ *Large four-battery aluminum flashlight with halogen bulb* — Not only does it allow you to peer into dark recesses, it lets you see the critters before they see you.

◆ *Pocket pad and pen* — To jot down those all-important notes and design inspirations you'll undoubtedly have when looking through your building.

◆ Are the sill boards in good condition?

◆ If the structure is brick or block, is the mortar crumbling?

◆ Are the bricks crumbling? Are any of the walls cracked?

Siding — Look at the siding closely both inside and outside the structure.

◆ Is the siding simply a covering, or does it help support the structure? (On many older buildings, the siding is an integral part of the building and cannot be removed without weakening the structure.)

◆ Are the nail heads pulling through the siding boards?

◆ Are the siding boards splitting?

◆ Are boards missing? Is daylight showing through the siding?

◆ Is any of the trim or wainscoting missing?

◆ Is the paint peeling? How many layers of paint do there seem to be?

Roofing — Once again, you want to look at the roof from both sides. Get a ladder and climb up to the eaves so you can see the condition of the roofing close up.

◆ What kind of roofing does the building have — shake, shingle, tin, or tile?

◆ If shake or shingle, how many layers are there?

◆ Is the roof sheathed, or are there simply stringers for the shingles?

◆ Is the flashing deteriorated or missing?

◆ Do the rafters show evidence of leaks or dry rot?

◆ Is the roof properly ventilated?

◆ Are the overhangs and the eaves in good condition?

◆ Is the guttering deteriorated or missing?

Doors and windows — Open and close all the doors and windows. Carefully inspect the hinges and hardware. On the windows, check the mullions and the glazing.

◆ Do the doors and windows open and close easily?

◆ Do the latches catch easily and are they secure?

◆ Are the doors and windows reasonably airtight? Do they let weather into the barn?

◆ Are the glazing and mullions in good condition?

◆ Are any panes missing?

◆ Do you need more or different doors and windows? How many?

◆ Do you need to change or add to the structure to mount modern door and window units?

Utilities — Make sure you turn off the power before you go poking around.

◆ Is the electrical wiring up to date? (This is one element you don't want to restore with old materials and techniques. Make sure the wiring is state of the art.)

◆ Is the service sufficient for what you want to do in the outbuilding? Do you need a sub-box or a larger service?

◆ Is there water or plumbing in the building? Do you need it?

◆ Is the drainage sufficient?

◆ Is there gas service to the building? Do you need it?

◆ Do you need a telephone in the building?

Foundations

When restoring an outbuilding, it's best to start at the bottom, even if the building is already standing. Unless you get the foundation squared away, you risk having to do the repairs you make to the rest of the structure over again. This can be intimidating — I know that for a fact. You have to screw your courage to the sticking-place to dig away under a standing building. However, it's not as difficult as it might seem. Break the job down into small tasks, and it all falls into place, so to speak.

Foundation problems normally fall into three categories, determined by the magnitude of the work you must do to put things back to rights:

1. The foundation is adequate, but it needs some repair. Stones may be loose, mortar may be crumbling, and the foundation may have settled unevenly, but all you have to do is repair or replace what's already there. You don't have to add anything new.

2. The foundation is inadequate, but what's there is serviceable. All you have to do is reinforce it or replace a portion of it.

3. The foundation is hopelessly inadequate or deteriorated, and it should be removed and replaced with a new one. This obviously requires the most work.

Minor repairs — fixing what's already in place

If the gods have smiled on you, the foundation of your building is sound and just needs a little attention. The guidelines that follow will help you to diagnose and fix minor problems.

The mortar is crumbling

If the mortar between the stones or blocks is crumbling away, it needs to be repointed. Don't put off this chore; a crumbling mortar seam allows water to enter the foundation. If this water freezes, it will expand, crumbling more mortar and loosening the masonry. In short order, the stones or blocks will begin to fall out of the foundation, and the problem will become more serious.

Clean out the cracks where the mortar is loose using a cold chisel and a mason's hammer. Remove all the loose mortar, making the crack deep enough to create a "key" for the new mortar to sit in. (*See Figure 3-1.*)

3-1. Using a long cold chisel, cut away the crumbling mortar between the stones or blocks. Stop when you reach solid mortar.

After you remove all loose mortar by hand, brush the joint clean with a wire brush, then flush the joint with water. Apply new mortar, forcing it into the crack with a striking tool. Use a pointing tool to smooth and surface the wet mortar. *(See Figure 3-2.)*

3-2. Use a pointing tool to smooth the wet mortar and blend it into the old mortar.

The concrete is crumbling

If one section of concrete is crumbling, check it all. It's possible that some long-ago builder poured a "weak mix" without enough cement to properly bind the sand and gravel. Poke and prod with a crowbar to determine whether the entire foundation is falling apart or you just have a bad section.

If you have a bad section of the foundation, excavate the crumbling concrete so that only the solid stuff remains. To tie the new concrete to the old, drill holes ¾ inch wide and 6 inches deep in the exposed face. Epoxy 18-inch lengths of ½-inch-diameter rebar in the holes. The rule of thumb is to use one piece of rebar every foot. Sink each bar about 6 inches into the old concrete and let it protrude about 1 foot into the area where you'll make the new pour. *(See Figure 3-3.)* Build a plywood form flush with the face of the old concrete and brace it so that it won't bow out when you make the pour. Pour the new concrete and let it cure for a day before you remove the form.

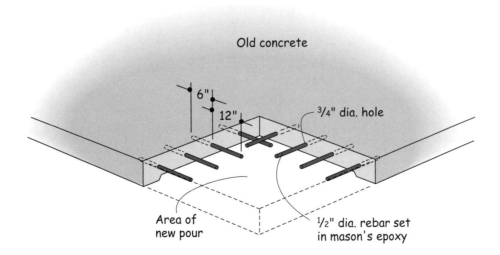

Old concrete

6"

12"

¾" dia. hole

Area of new pour

½" dia. rebar set in mason's epoxy

3-3. Tie the old concrete to the new pour with lengths of ½-inch-diameter rebar. Using a masonry bit, drill ¾-inch-diameter holes in the old concrete, 6 inches deep. Secure 18-inch-long pieces of rebar in the holes with mason's epoxy or patching cement.

3-4. To quickly remove the old mortar from a brick or stone, use the pane (pointed end) of a mason's hammer like a chisel.

The stones or blocks are loose

If your foundation wall is made of stones or bricks and some of these are loose or missing, you must replace them. The procedure is similar to that for pointing the mortar. Remove the loose pieces and chisel away all the soft mortar with a hammer and a cold chisel. Brush the dust from the cavity and flush it out. Also clean the loose masonry, removing all the old mortar with a mason's hammer. *(See Figure 3-4.)* Using a trowel, spread a layer of mortar on the top surfaces of the old stones or bricks that are still in the wall. Also spread mortar on the top surface of the loose or replacement masonry. Slide the new bricks or stones into place and smooth the mortar with a pointing tool.

A portion must be reinforced or replaced

Frequently, old barns and outbuildings have inadequate foundations. If the structure isn't adequately supported, several problems can develop. The building may settle unevenly or slip off the foundation. The foundation may crack, split, or crumble because the weight of the structure is too great. The remedy for all these problems is to replace the damaged portion of the foundation, then add to or reinforce the foundation walls so that they provide the proper support.

EXTRA HELP

When you are pointing old mortar or replacing loose masonry, the amount of water you use in mixing the mortar is very important. The mortar mustn't be too stiff or too runny. It should be "sticky." Test the mix by wiping some mortar on the face of a brick or stone. It should be wet enough to stick to the masonry for several seconds but dry enough to hold its shape without running off.

3-5. Portions of this barn have slipped off its foundation, and other parts have settled unevenly, because the foundation does not provide enough support.

The foundation has settled unevenly

Before you do anything, determine why the foundation has settled unevenly. Do the foundation walls extend below the frost line? Is water being properly drained away from the foundation? Does the building rest on ground with uneven soil composition? Is there an abandoned well or cistern near the foundation wall?

If the foundation walls don't go deep enough, or if water isn't drained away from the foundation, the structure has likely suffered from frost heaves. As the water in the soil freezes in wintertime, it expands and heaves up. This will lift a building, even a large one. And since the ground won't thaw evenly, the building won't settle evenly. Over many seasons, this unevenness will become more and more pronounced.

Water in the soil may have a similar effect. If the water isn't drained away from the structure, the soil will remain saturated. As it dries out, it won't dry evenly: Parts remain soft and muddy, while others dry and harden. The difference in saturation lets the building settle unevenly. Once again, the unevenness isn't immediately apparent. It becomes evident only after the building has stood for many years. *(See Figure 3-6.)*

If the soil is uneven in its composition, the foundation will settle unevenly. The carriage barn behind my house is just such a case. The north side of the barn faces an alley, where street crews have applied gravel for many years. For decades, horses and then cars have compacted the mix of soil and gravel so that it's almost as hard as rock. Behind the barn, on the south side, is a large garden. Generations of owners have applied compost and worked the soil so that it's soft and loamy. Not surprisingly, the barn tilted decidedly toward the garden when I first moved in.

3-6. The deep rut running parallel to the foundation wall of this barn is a sure sign that water is not being drained away from the foundation. This is where the runoff from the roof hits the ground. Gutters and downspouts will fix the problem.

3-7. *The well next to this old corn crib is slowly collapsing. The ground has shifted, and the crib has slid off the foundation. Before you jack the barn up and put it back on its foundation, you must fill in the well to prevent the ground from shifting further.*

However, one corner was much lower than all the others. After a little digging, I discovered an old, unused cistern near this corner. It probably once collected rainwater to irrigate the garden. When the cistern was allowed to go dry, the walls collapsed. The earth around the cistern shifted, and the corner of the barn shifted with it. (*See Figure 3-7.*)

What do you do in cases like this? The first step is to correct the situation that caused the uneven settling, if you can. If the barn foundation has suffered because the surrounding soil is improperly drained, some guttering, downspouts, and drainage tiles will usually fix the problem. If a cistern, retaining wall, or another structure nearby has allowed the ground to shift, fill in the cistern or repair the retaining wall.

The action you must take to stop the foundation from moving depends on the cause of that movement.

Foundation walls are above the frost line — To fix this problem, you must extend the foundation further down into the ground. This is a time-consuming operation, but not a difficult one. You can remove small sections of the foundation wall at a time and replace them with deeper sections. (*See Figure 3-8.*) Or, if you don't want to replace the entire foundation, insert deep piers in the foundation wall at key spots. (*See Figure 3-9.*) In a post-and-beam structure, these piers should be placed under the posts. In a frame or log structure, the piers should be spaced 6 to 10 feet apart, depending on the size of the structure. The larger the structure, the closer the piers should be.

To pour a pier or replace a section of the foundation wall, first brace the frame where you will be working and jack up the structure so that it's off the foundation in the work area. (See page 60 for instructions on how to lift a building, or a portion of a building, off its foundation.)

Carefully remove a section of the old foundation wall. Don't remove too much at a time! Dig down below the frost line and pour a concrete footer. Build the wall section or the pier on top of the footer. You can lay up the wall or the pier from stone if you want to preserve the old-time look of the structure. If expedience is more important than aesthetics, lay up the wall or pier from concrete block. Or, build a wooden form and pour a concrete wall or pier. Whatever method you choose, take care that the tops of the wall section or the piers are absolutely level with one another. When you let the barn back down on the reinforced foundation, it should rest level once again.

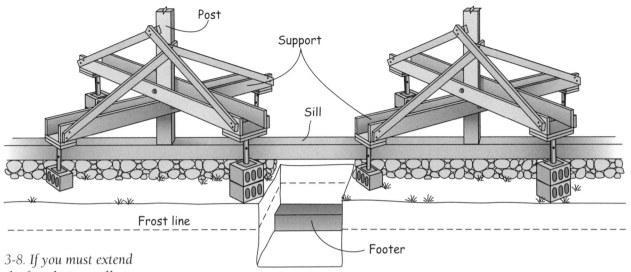

Post

Support

Sill

Frost line

Footer

3-8. If you must extend the foundation wall below the frost line, brace the structure, jack up one portion, remove a 4-foot-long section, and replace it. Then build the wall one new section at a time; don't try to pour a footer under the existing wall.

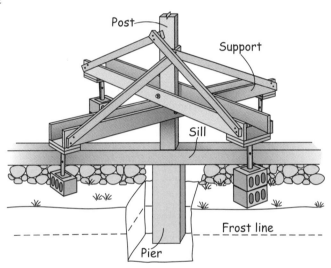

Post

Support

Sill

Frost line

Pier

3-9. You can also remove small sections of the old foundation and replace them with deep piers. This effectively converts a standard wall foundation to a pier-and-beam foundation.

3-10. The foundation of this carriage barn was replaced a section at a time. I removed an 1820s vintage stone foundation that was barely a foot deep in the soil and replaced it with a footer and concrete block wall extending down 3 feet.

The building rests on soil of an uneven composition — The remedy is the same as that for a foundation that rests above the frost line. Extend the foundation, making the walls or piers go deeper. The deeper the foundation reaches into the earth, the less the density of the surrounding top soil will affect the stability of the structure.

The foundation is improperly drained — If your building "takes on water" every time it rains, there are several possible solutions.

◆ Grade the surrounding ground so that it slopes away from the building. The water will then run downhill, away from the foundation.

◆ Install tile drains at the bottom of the foundation. Even century-old stone foundation walls were sometimes laid with clay tile drains. These tiles usually direct the runoff into a cistern close to the building, or they take the water down slope and away from the building. It's quite possible that when you dig down to lay new tile you'll find the remains of the old tile that, after years of use, have either filled up with silt or been crushed.

◆ If your building has a basement, excavate a 1-foot-wide, 1-foot-deep trench around the perimeter and install a drainage hose connected to a sump pump. The pump will remove water from the "sump pit," keeping water away from your basement floor and the foundation.

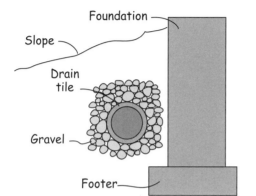

3-11. *To properly drain the foundation, place the drainage tile near the footer. When you fill in the trench you dug for the tile, grade the soil so that it slopes away from the building.*

EXTRA HELP

Without a doubt, the best tool for checking that a foundation is level is a garden hose. Lash or tape one end of the hose to the pier or wall section that you just built so that the end is flush with the top surface. Attach the other end to a stake next to the pier or wall section you are building. Fill the hose with water so that the liquid overflows the first end. Adjust the level of the second end until water starts to overflow there, too. Build the next pier or wall section dead even with the second end of the hose, and it will be perfectly level with the last pier or wall.

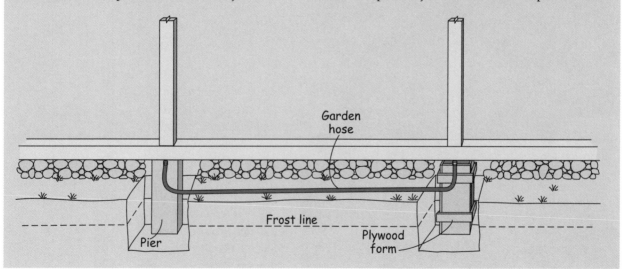

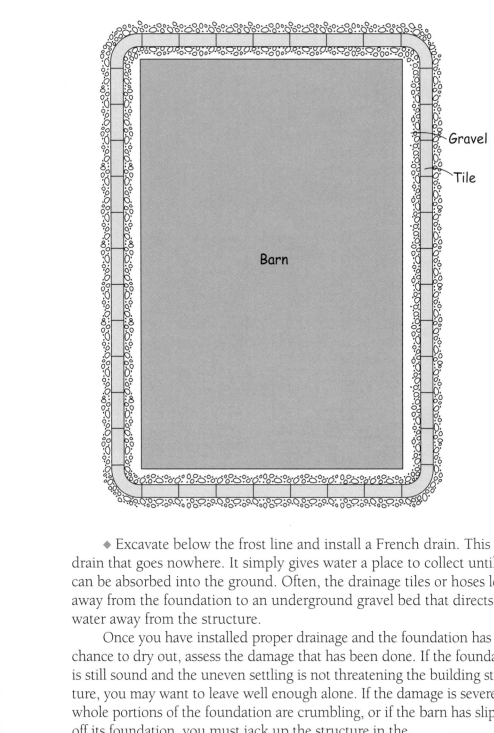

Gravel

Tile

Barn

3-12. Drainage tiles should ring the perimeter of the building. If you want to lead water away from the building, dig a 3-foot-deep trench out from one corner and fill it with 2 feet of gravel topped with 1 foot of soil. Use washed "pea" gravel ½ to 1 inch in diameter.

◆ Excavate below the frost line and install a French drain. This is a drain that goes nowhere. It simply gives water a place to collect until it can be absorbed into the ground. Often, the drainage tiles or hoses lead away from the foundation to an underground gravel bed that directs water away from the structure.

Once you have installed proper drainage and the foundation has had a chance to dry out, assess the damage that has been done. If the foundation is still sound and the uneven settling is not threatening the building structure, you may want to leave well enough alone. If the damage is severe and whole portions of the foundation are crumbling, or if the barn has slipped off its foundation, you must jack up the structure in the vicinity of the damage, remove the damaged area, and repair or replace it.

A nearby well or cistern is collapsing — Fill in the cistern or the well with rock and soil to stop the collapse. This will prevent the foundation from settling any further. Again, once the root of the problem has been identified and corrected, assess the actual damage to the foundation. If the damage is severe or threatens to become so, repair the foundation.

EXTRA HELP

If you expose the walls of your foundation when you install a drain or perform some other repair, take the opportunity to clean the walls and coat them with waterproofing material to help protect them. This process is called *parging*.

Raising a building

Many foundation repairs require that you raise the building. No, I don't mean raising in the old-time sense of erecting a structure; I mean lifting it off its foundation. Generally, all you have to do is lift the building a few inches and hold it there while you either repair or replace the foundation. This operation requires several inexpensive floor jacks, some lumber, and a good deal of hard work.

1 To lift a frame-constructed building, make a *lifting frame* and attach it to the studs. Nail a 10-foot-long horizontal 2x8 beam to each stud, then brace it with 2x4s, as shown. Tie the ends of the beams together with 2x4 stringers. To do this, you will have to remove some of the siding from the building. For clarity, the illustrations show all the siding removed. In reality, you have to remove only the siding where the beams and the braces attach to the studs.

Tamp the earth beneath the ends of the beams to compact the soil, then place a concrete block beneath each end. Place floor jacks between the beams and the blocks. Slowly raise each jack an inch at a time, working in a circle around the lifting frame until the building frame is suspended the desired distance above the foundation. For most repairs, you have to lift the building only a few inches free of the foundation.

2 You really need only one or two jacks to get the job done. Install the block and jack 1 foot in from the end of the beam. Stack concrete blocks up under the end. When you can't fit any more blocks under the beam, stack 2x4 scraps. Use the jack to raise the beam high enough to slip in another 2x4 scrap. Let the beam down on the scrap, move the jack to another beam, and repeat. Once the stack of scraps approaches 8 inches tall, remove it and slip another block under the beam.

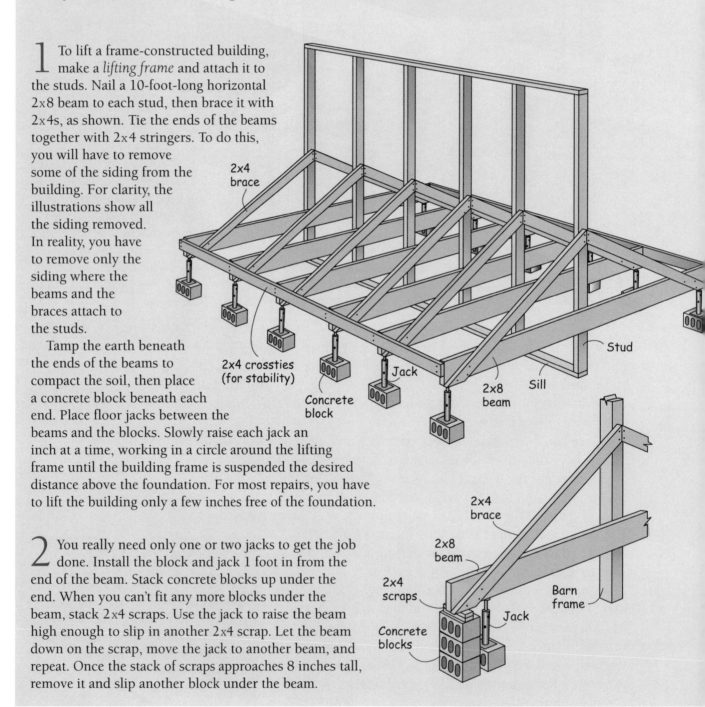

2x4 brace

2x4 crossties (for stability)

Concrete block

Jack

2x8 beam

Sill

Stud

2x4 brace

2x8 beam

2x4 scraps

Concrete blocks

Jack

Barn frame

3 What if you need to raise a post-and-beam building? The procedure is similar, but the lifting frame is somewhat different. Attach a 12-foot-long horizontal 2x8 beam to each corner of the post you want to lift, as shown. Secure the beams to the posts with ½-inch-diameter lag screws, then brace them with 2x4s. Tie the beams together with double 2x8 crossties nailed to the bottom edge of the beams.

Place concrete blocks under the ends of the beams and insert floor jacks between the beams and the blocks. Raise each jack in turn an inch at a time until the post comes off the foundation or the sill.

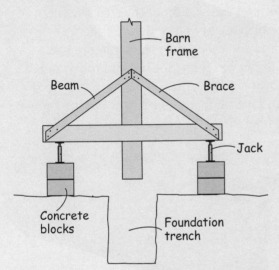

Lag screws

Post

2x4 braces

2x8 beams

Post

2x8 crossties (doubled)

Concrete block

Barn frame

Beam

Brace

Jack

Concrete blocks

Foundation trench

4 Once you have lifted the structure, you can begin foundation repairs or remove the old foundation and install a new one. When digging a foundation trench, keep it as narrow as possible.

5 Be aware that the weight of the building on the concrete blocks may cause the trench to collapse, particularly if the soil is soft, is sandy, or becomes saturated with rainwater. To prevent this, line the trench with 2x10s as you dig down. Brace the 2x10s against the trench sides with 2x4s and wedges, as shown.

You can pour a concrete footer directly under the 2x4 braces without removing them, if the work calls for it. Remove the 2x4s and backfill the trench as you build up the foundation wall.

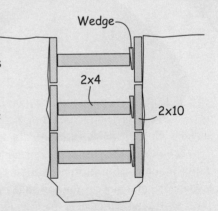

Wedge

2x4

2x10

TO BE SAFE

Brace the structure every which way from Sunday before you lift it. Don't lift it all at once; just raise a bent or a section at a time. If you must lift it all at once, hire or consult a professional building mover. Also consider disassembling the structure, and then putting it back up again. This may be the safest, wisest, and fastest course, particularly if you intend to replace the siding and the roof.

The structure is slipping off its foundation

You may notice a large bow in the side of your building and, may upon further inspection, find that a section of wall has slipped off the foundation. Uneven settling, a collapsing cistern, and other causes I've mentioned previously may cause this problem. It can also be caused by an inadequate foundation. Older barns were often constructed with posts simply resting on stones. Over time, the natural movement of the wood as it shrinks and swells with changes in humidity causes the post to work its way off the stone. Even a whole wall section can "walk" off a stone wall foundation.

The thing to do is to replace the stone with something more substantial and tie the barn structure to it. First, brace the structure in the area where you will be working. Then lift it up off the foundation stones, remove the old foundation in this area, and excavate for a new one. Build or pour a wall or a pier directly under the section of the structure that you've lifted. Set anchor bolts or post anchors in the top of the wall or the pier. Let the mortar or the concrete set, then lower the structure back onto the improved foundation. If you've set anchor bolts in the foundation, drill holes in the sill to fit over the bolts. Secure the structure to the foundation by tightening nuts and washers on the bolts or by driving nails and screws through the post anchor. (See Figure 3-13.)

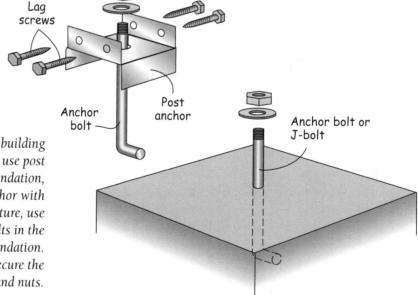

3-13. To secure a post-and-beam building or a pole building to a foundation, use post anchors. Bolt the anchor to the foundation, then attach the post to the anchor with nails or screws. For a frame structure, use anchor bolts. Set the anchor bolts in the wet mortar or concrete of the foundation. Fit the sill over the bolts and secure the structure with washers and nuts.

The poles are rotting in the ground

If you have an older pole building, the poles may be rotting in the ground. Even the best chemical wood preservatives can't keep wood in ground contact from rotting eventually. To save the building, you must cut off the poles before the rot spreads up past the splashboards. Rest the cutoff ends of the poles on concrete piers.

The procedure is almost exactly the same as putting a building back on a foundation: Raise the structure one pole at a time. As you take the weight of the building off a pole with a lifting frame, cut the pole off even with the bottom edge of the splashboards or the bottom girts. (*See Figure 3-14.*) Excavate for a pier, removing the rotted end of the pole from the ground. Pour a concrete pier and set a post anchor in the top. After the concrete has set, lower the structure onto the pier and secure the pole to the anchor.

The building is slipping down a hillside

There is no easy fix for a building that's falling down a hill, so consider whether the building is worth saving. The solution to this problem is to stabilize the earth below the building — in other words, keep the hill from eroding. (*See Figure 3-15.*) It may require a retaining wall; it may require something more. The first thing you should do is consult a geologist in your area. The county extension office is a good place to start looking for this person.

The geologist will bring up a new set of concerns. One of the first questions he or she will ask is, "How fast is the building moving?" If you just purchased the land, you'll have no way of knowing. Even if you've had the land for some time, it's not something that you usually measure. The geologist may suggest that you shoot the building with a transit, then shoot it again six months later to determine precisely how fast the building is moving. If you're lucky, you may find that the building has stabilized. The ground may have been in motion in the past, maybe during particularly rainy weather, but is currently stable. If this is the case, you can proceed with normal foundation repairs.

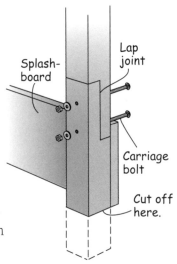

Splash-board
Lap joint
Carriage bolt
Cut off here.

3-14. Cut off the rotten poles even with the splashboards. If the rot has spread further up the pole, remove the rotted portion and splice in new, sound wood.

3-15. This barn seems to be falling down a hill, but it's not so. The hill is actually dragging the barn with it as it erodes.

3-16. These stone boxes — large stones in a wire cage — are stacked like huge bricks to create a retaining wall that stabilizes this hillside.

If you determine that the building is definitely slipping, find out how much. I had a friend whose building was slipping down a hill because the area below his building had been timbered and the tree root systems that were holding the hill together were no longer there. He was able to stabilize the hill by planting a fast-growing species of pine that stabilized the hill in five years. His building, thankfully, was moving less than an inch per year, so he had time to let the trees grow. Once the hill was stabilized, he was able to repair the foundation, so the building is standing straight today.

If you don't have time to plant trees, or if trees don't grow well in your area, stabilize the hill by taking the "DOT" approach. Often the Department of Transportation anchors long concrete beams into a hill to create a concrete retaining wall. The beams hold the retaining wall in place. They also keep the wall from bowing out under the weight of the earth behind them. This stabilizes the earth above the wall and keeps it from sliding down. You can also purchase "stone boxes" made from galvanized fencing and field stone. Dig into the hill and stack these boxes (with the aid of a crane) to create a retaining wall that will prevent further erosion of the hillside. *(See Figure 3-16.)* These approaches are not cheap or easy, but they will help to save the building.

Major repairs — replacing the foundation

Many times, you'll see a building that has been built on an old foundation that wasn't designed to take the weight of the new structure. It is also common to find buildings that were built with inadequate foundations — our forefathers weren't always the best of engineers. Walk around the building and you'll see the tell-tale signs: crumbling walls, loose and cracked masonry not just in a few spots, but everywhere. *(See Figure 3-17.)* You may even find that your building rests on a few stones or sits right on the dirt! It was once common practice to erect a crib barn on oak sill logs resting right on the earth.

When the foundation is disintegrating and wood is in direct contact with the ground, you can't simply repair the foundation or reinforce it. You must completely replace it. You can do this with the building still standing. It's time-consuming, but not particularly difficult to jack up the building one section at a time and build a foundation beneath it. However, give some serious thought to disassembling the building to do the foundation work, especially if you have a post-and-beam barn or a log structure. It's a lesser job to take these buildings down (especially one as big as a barn) than to jack them up. Remove the roof and the siding and knock out any pegs. Carefully label the parts as you disassemble them so you can put everything back together in the proper order.

If you're restoring a building that has been nailed together, that's a different story. These buildings are much harder to disassemble, and it's a safe bet that you will damage some of the boards in the process. Furthermore, there are too many parts to label. If the structure is sound, it's much easier to jack it up and install a foundation beneath it.

A foundation primer

If you must replace a foundation, you should first decide what kind of foundation you're going to build. And before you can make a good decision, you must understand that a foundation serves five purposes:

1. A foundation carries and distributes the weight of the structure to solid earth.

2. It provides a level and stable base for your building.

3. It keeps the building from settling into ground.

4. It acts as an anchor for the building against high wind and seismic forces.

5. It keeps moisture from affecting the building.

There are five types of foundations that are commonly installed beneath barns and outbuildings.

A *perimeter foundation* is a wall that sits on a footing. A large perimeter wall could be a basement wall. The footing is wide and deep enough to carry the load of the perimeter wall. The perimeter wall can be made from just about any stone-based material, or even treated lumber. It can be laid up using brick or block or formed up with wood or metal forms and poured. It can also be poured using new permanent foam forms that act as basement wall insulation.

3-17. *This foundation is crumbling not because it's old but because it can't support the weight of the building above it.*

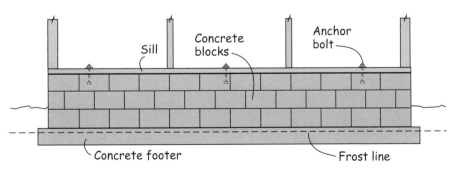

3-18. *Perimeter foundation*

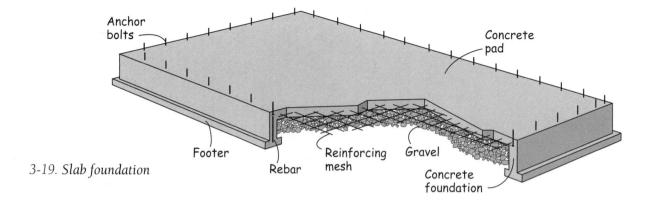

3-19. Slab foundation

A *slab foundation* is just as it sounds. Pour a slab over a footer and wall. The footer, wall, and slab are usually poured all at once (this is known as a *monolithic slab*). But that's not possible if you're installing a slab under a standing building. You must pour the footer and walls first, then the slab. The footer should rest below the frost line. Place the slab over a bed of gravel and reinforce it with wire mesh.

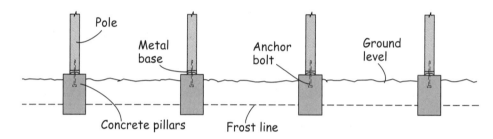

3-20. Pier foundation

A *pier foundation* consists of footers and piers. Each footer must be large enough to support the pier and keep it from sinking into the earth. The number and spacing of the piers will depend on the size and weight of the structure they support. Pour the footers below the frost line, setting in rebars to help tie them to the piers. Set the forms for the piers on the footer. You can make square forms from plywood or use ready-made sonotubes. Pour the piers on top of the footers.

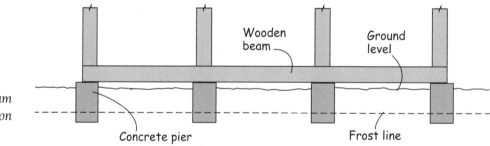

3-21. Pier-and-beam foundation

A *pier-and-beam foundation* consists of piers tied together by a long beam. The beam can be poured out of concrete, but it's more often a wooden beam. The beam distributes the weight of the building across the piers. Pour the footers and piers exactly as you would for a pier foundation, then lay a beam across them.

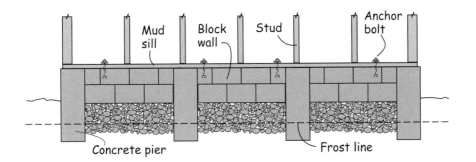

Mud sill   Block wall   Stud   Anchor bolt

Concrete pier   Frost line

3-22. Floating foundation

A *floating foundation* combines features of a pier-and-beam and a perimeter foundation. Space concrete piers every so often along the walls and tie them together with a perimeter wall. This wall floats on a gravel bed, allowing slight movement. To make the wall, lay cement blocks on top of each other without mortar between each block, oriented so that the voids in the block run through the wall from top to bottom. Fill the voids with cement to make a solid wall and allow it to cure.

If you're not going to disassemble the building to install a foundation, you'll have to jack it up. You don't have to lift the entire building all at once. For most of the foundations I have mentioned, you can raise one section at a time and install the foundation as you go, as I explain on page 60. The exception to this rule is a slab foundation. In this case, the entire building must be suspended while you set the forms and pour the concrete. If you must go this route, get help from a professional building mover.

EXTRA HELP

If you want your restored building to have a concrete slab floor, you don't have to pour a slab foundation. You can build a perimeter foundation or a floating foundation, then pour the slab inside the walls. *See page 83.*

Marrying the building to the foundation

Whatever foundation you build, remember that you must tie the structure to it. There are several ways to do this. The most common tools for the job are anchor bolts (also called *J-bolts*), lead anchors, hurricane straps, and post anchors (also called *buckets*).

If you use *anchor bolts*, insert them into the wet concrete or mortar before it hardens. Let the bolts protrude from the top of the foundation wall or pier. They must extend up through the sill plate far enough so that you can thread nuts on the ends of the bolts. (*See Figure 3-23.*)

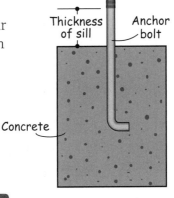

Thickness of sill   Anchor bolt

Concrete

3-23. Anchor bolts

Lead anchors (commonly called *red heads*) can be installed after the foundation wall has cured. Drill holes through the sill plate and into the foundation with a masonry bit. Drop the lead sheath into the hole, then drive a lag screw into the sheath. The sheath will expand in the concrete, wedging itself in the hole, and secure the sill to the foundation.

Hurricane straps can be used to tie a building to a foundation or to tie a roof to a building. They are metal straps that keep the joined surfaces from pulling apart in a storm. Install them using hardened nails or screws. (Hardened fasteners are less likely to shear off.)

Use *buckets* to attach sill plates, posts, and poles to concrete and masonry piers. Bolt the buckets to the masonry with anchor bolts or lead anchors, then nail the flanges on the buckets to the wood.

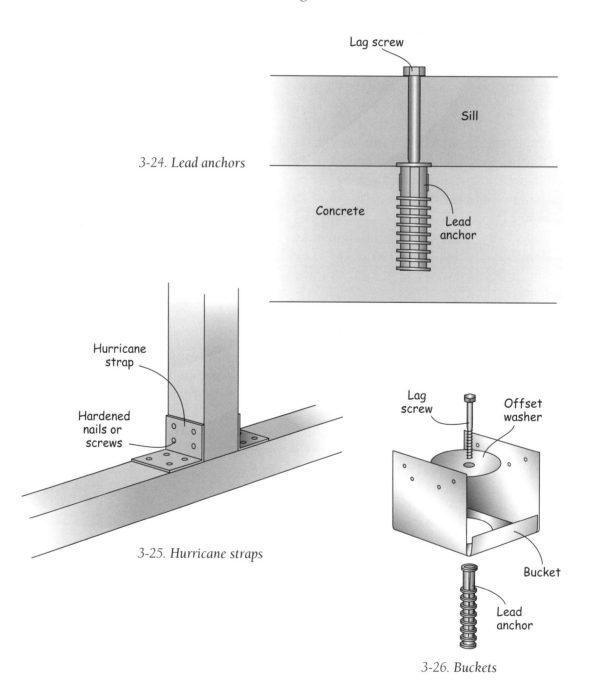

3-24. Lead anchors

3-25. Hurricane straps

3-26. Buckets

Floors

I have yet to renovate a barn or other outbuilding in which the floor didn't need major repairs or replacement. Mice chew their way through floorboards, moisture rots joists, and cycles of freezing and thawing cause concrete slabs to crack. On top of that, the floor gets more wear and tear than any other part of a building.

If your renovation project includes floor repairs, give some careful thought to the type of floor you need in the building. It may not be adequate to restore the existing floor, especially if the purpose of the building will change. More often than not, you'll need to upgrade the floor or change it completely. For example, if you are restoring an old storage structure with a concrete floor to use as a workshop or office space, consider tearing up the concrete floor and replacing it with wood — wood is much easier on your feet. If you want to use the building to store heavy machinery, then you may have to rip up the concrete floor and replace it with thicker concrete that is reinforced with girders.

Finding and fixing flooring problems

Weak, sagging, cracked, and deteriorating floors are often symptoms of other structural problems. You must correct these deeper problems before you repair or replace the floor; otherwise you'll just have to do the work over in a few years. In Chapter 3, I described a situation in which dirt and debris were allowed to accumulate over the years around the perimeter of a building, raising the level of the ground above the foundation so that the wood siding was in contact with the ground. Now suppose this structure had a wood floor system vented by a crawl space underneath with small vents in the foundation wall. If a wood floor doesn't have air moving under it to prevent moisture from accumulating, it will eventually rot. If the debris and dirt cover the vents, the moisture content in the floor joists will rise, allowing insects and bacteria to eat away at the wood. The next thing you know, your restored barn floor is sawdust! Don't just fix the flooring problems; fix the causes of the flooring problems.

4-1. Dips in a floor are usually caused by problems with the joists supporting it or with the foundation holding up the joists.

The floor is weak, uneven, sagging, or sloped

Weak and sagging wood floors are usually caused by a structural failure — the joists are rotting, or the floor has been stressed by more weight than it can safely support. In either case, you must repair or replace the deteriorating structural members and possibly shore up the floor.

If the floor is uneven or sloped, the problem probably begins with the foundation. You must repair or replace the foundation first, then turn your attention to the floor.

Find out how and where the floor departs from level. Stretch string from corner to corner and wall to wall to determine where the floor is high and low. (*See Figure 4-2.*) Use a marble to find how the floor is sloped; the marble will roll to the lowest point. Also, stand at various places on the floor and jump up and down. If the floor bounces, it's weak at that point.

If you have a wood floor, inspect the joists, especially where the floor is low. Are they rotted or broken? Do they seem to be curved or bowed? What size are they? Are they wide enough and thick enough to support the weight on the floor? What materials comprise the existing flooring?

If you have a concrete floor, drill a small hole through a low part with a masonry bit. How thick is the flooring? Is it properly supported and drained? Did the drill bring up any gravel or sand beneath the floor? Are there expansion joints in the slab or around the perimeter? Where parts of the floor have broken away, do you see any evidence of reinforcing mesh or rebars?

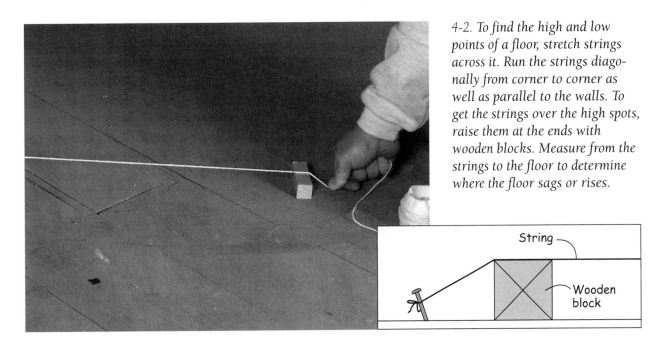

4-2. To find the high and low points of a floor, stretch strings across it. Run the strings diagonally from corner to corner as well as parallel to the walls. To get the strings over the high spots, raise them at the ends with wooden blocks. Measure from the strings to the floor to determine where the floor sags or rises.

String

Wooden block

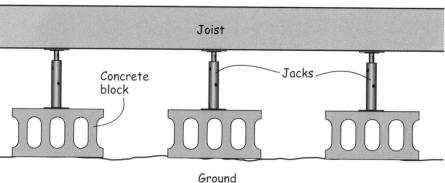

4-3. Place a large, flat concrete block directly beneath the joist to provide a foundation for the jack. Measure the distance between the block and the joist, then choose a jack with a body that's a few inches shorter. (Some jack bodies are adjustable, allowing you to change their height to fit your needs.) Place the base of the jack on the concrete block. Raise the screw by hand until it bears against the joist, then turn the screw with a wrench until the joist is at the proper height.

Installing floor jacks — You can shore up a weak or sagging wood floor by installing *floor jacks* beneath the joists. Floor jacks are inexpensive screw jacks made just for this purpose. They come in several lengths to accommodate the height of your crawl space or basement. (See Figure 4-3.)

Place the jacks under the low spots in the floor, then raise them to make the floor more or less level. In cases where you need to jack up several neighboring joists, make a thick beam that will run perpendicular to the joists, spanning them from below. Place several jacks under the beam, then raise the beam with the jacks so it presses up against the joists. This not only saves you the expense of buying a jack for each joist, it also helps keep the joists at the same level. (See Figure 4-4.)

E X T R A   H E L P

Use hanger straps (available in the plumbing section of your hardware store) to hang the beam from the joists while you position the jacks.

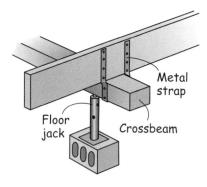

4-4. When you must raise several neighboring joists, make a beam to span them. Using hanger straps, hang the beam perpendicular to the joists. Place floor jacks under the beam and raise the jacks so they press against the beam and lift the joists.

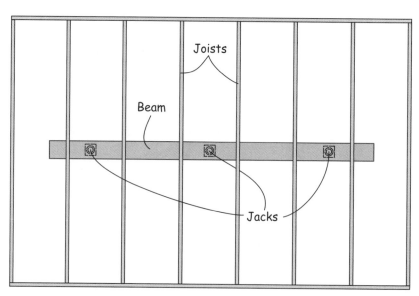

Raise the jacks slowly *in turn*. Turn each jack just one or two times, then go on to the next. Repeat in the same order each time you raise the jack. Some portions of the floor will come up to the level at which you want them before others do. As this happens, stop turning the jacks under that portion. Continue turning the other jacks a little at a time and check carefully until the floor is as level as you can make it.

Leave the jacks in place permanently. Check them monthly for the first year if the cement blocks rest on bare earth. The blocks will settle, and you will have to raise the jacks slightly until the settling halts.

Installing bridges — One way that you can strengthen a weak wood floor considerably is by installing *bridges* between the joists. As their name implies, bridges connect the joists so that a load on any one part of the floor is distributed across the entire floor. Install the bridges perpendicular to the joists, every 10 feet or so along their span. (*See Figures 4-5 and 4-6.*)

4-5. The bridges between the joists create trusses that connect all parts of the floor. These distribute the load resting on the floor across a wider area, making the floor stronger and more solid.

4-6. There are several types of bridges that you can install. For simplicity, cut short lengths of 2X lumber and nail them between the joists. Note that you must stagger these blocks so you can drive nails into the ends. Alternatively, cut strips of lumber and miter the ends so they form an X-shaped brace between each set of joists. You can also purchase metal strips that are designed to be installed as X-braces. To use these metal braces, however, the joists must be a standard 16 inches apart, on center. The spacing between joists in old buildings often varies considerably.

Block

Wood X-brace

Metal X-brace

Shimming joists — The frames of older structures were often put up "green." The wood was cut soon after the log was felled, so the lumber had a high moisture content because much of the sap remained. There were two reasons for using green wood. First, it was much easier to drill and nail, especially in the building of frames from hardwoods, such as oak. Second, it was thought that the green lumber would shrink around the nails and pegs as it dried, helping to hold the joints tight.

Unfortunately, wood not only shrinks as it dries, it also tends to bow and twist. As a result, some of the joists in your structure may be far from straight, and the floor that is laid over them will be far from level.

To fix this problem, you must remove the flooring to expose the joists. Along each wall, snap level lines to run at the height where you want the floor to rest, and stretch strings beside the joists. *(See Figure 4-7.)* Where the joists are high, plane them down so the strings just barely touch them. To fill in the low spots, nail 2x4s to the sides of the joists so the top edge of each 2x4 is even with the strings.

Replace the flooring, nailing it to the 2x4s rather than the joists.

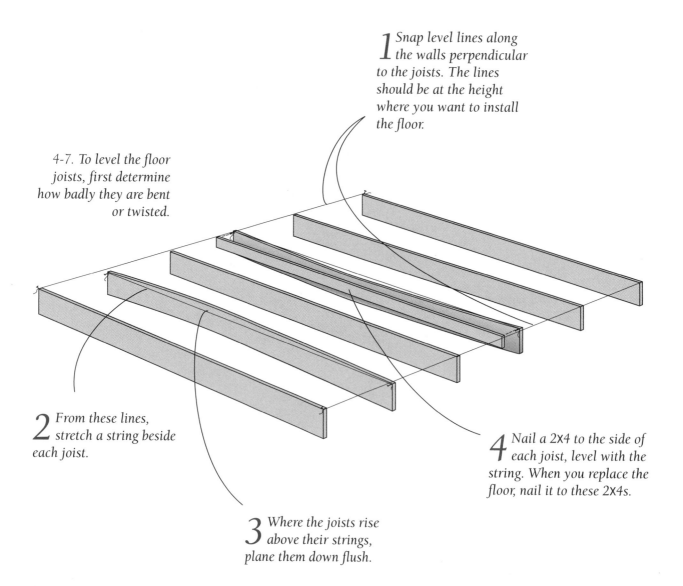

1 Snap level lines along the walls perpendicular to the joists. The lines should be at the height where you want to install the floor.

4-7. To level the floor joists, first determine how badly they are bent or twisted.

2 From these lines, stretch a string beside each joist.

4 Nail a 2x4 to the side of each joist, level with the string. When you replace the floor, nail it to these 2x4s.

3 Where the joists rise above their strings, plane them down flush.

Sistering joists — If the joists are rotted, cracked, or broken, you must "sister" them. To sister a joist, apply new wood on one or both sides of the old joist to span the damaged portion and reinforce it.

With the jacks ready, place the new joist section alongside the old one. The new wood must span the damaged area of the old joist so you can attach both ends of the new piece into solid wood. Attach one end of the new joist section to hold it in place while you position the jacks. Raise the old joist so that the floor above it is level. Carefully and slowly jack up the other end of the new joist section until it's parallel to the old joist. Be careful not to raise the joists too high and damage the flooring above. Attach the other end of the new joist section to the old, then drive fasteners at 12-inch intervals through the new wood and into the old. (See Figures 4-8 and 4-9.)

What if the end of a joist is rotted? Jack up the old joist to level the floor and slide a new sister section in place so that it rests on the sill or plate. Raise the sister section parallel to the old joist and secure it with lag screws.

EXTRA HELP

You can sometimes save a cracked joist without sistering it by driving the lag screws through the edge of the board to draw the crack back together.

4-8. When sistering a joist, make the sister section from wood that's the same thickness and width as the old joist. Cut it to span the damaged area of the joist with several feet to spare on either side, and attach it to the joist with lag screws or carriage bolts.

4-9. Using jacks, raise the old joist until the floor is level, then bring the sister section parallel to the old joist. Do this slowly and carefully — if you move the joist too quickly or too far, it may damage the floor above. When the joists and the sister section are parallel, secure the other end of the sister section to the joist and remove the jacks.

When a floor is sagging badly, several joists are usually giving way. This is especially true when the problem is caused by rot or insects. The conditions that affect one joist and cause it to rot will affect its neighbors as well.

When several joists are giving way, the first step is to fix the underlying problem. Reduce the moisture, ventilate the crawlspace, and kill the bugs. Next, cut and hang a beam that spans the damaged joists. Raise the beam with floor jacks until the floor above it is level. *(See Figure 4-10.)* Sister each one of the damaged joists, sliding sister sections over the beam.

You may find this difficult to do, especially in cramped quarters. The sister sections tend to get wedged between the beam and the flooring. To avoid this problem, rip a ¼ inch or so from the edge of the sister sections, making them slightly narrower than the joists. In extreme cases, I have ripped the sister sections in two and slid each half in place. *(See Figure 4-11.)*

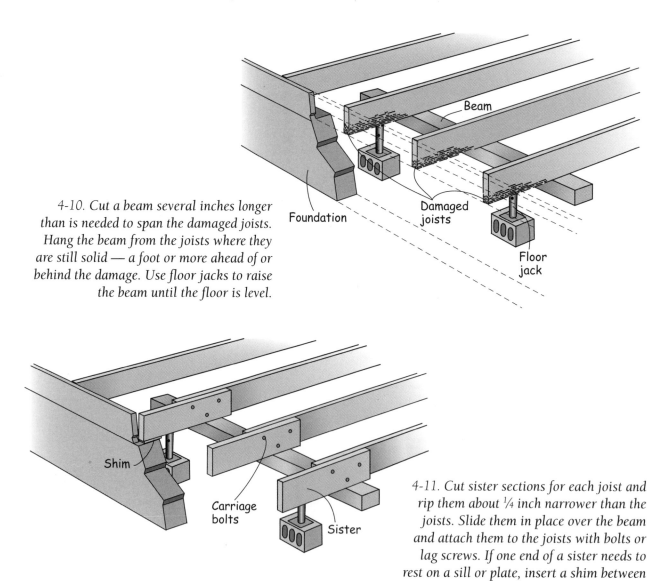

4-10. Cut a beam several inches longer than is needed to span the damaged joists. Hang the beam from the joists where they are still solid — a foot or more ahead of or behind the damage. Use floor jacks to raise the beam until the floor is level.

4-11. Cut sister sections for each joist and rip them about ¼ inch narrower than the joists. Slide them in place over the beam and attach them to the joists with bolts or lag screws. If one end of a sister needs to rest on a sill or plate, insert a shim between the sister and its supporting surface.

Framing fasteners

When performing repairs on floor joists and other sections of a frame construction, you'll find metal *framing fasteners* can help with all sorts of odd jobs. With a little imagination, these fasteners can be adapted to perform tasks beyond what they were designed to do. For example, a truss strap, designed to fasten roofing truss to top plates, is perfect for hanging reinforcing beams from damaged joists. Gussets and mending straps will hold splices together. T-plates and corner braces tie a supporting post to a joist or beam. All of these fasteners are designed to be nailed or screwed to the wooden frame members. When you purchase materials to do the repairs, take a stroll through the fastener section of your hardware store to see if anything there will make the job a little easier.

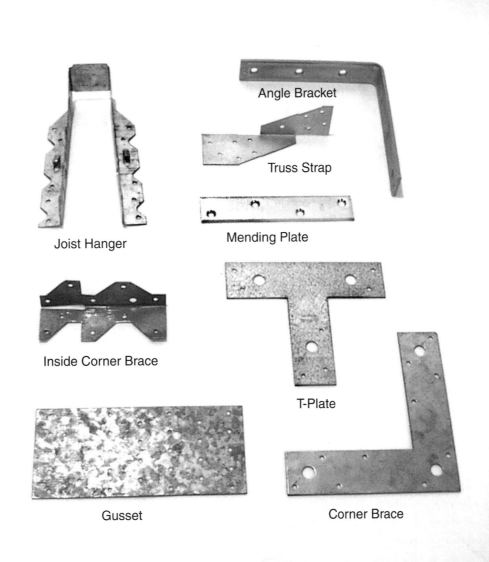

Joist Hanger

Angle Bracket

Truss Strap

Mending Plate

Inside Corner Brace

T-Plate

Gusset

Corner Brace

Leveling a concrete floor

If the building has a concrete floor that is uneven or sloped, it's probably best to excavate it and pour a new floor rather than salvage the old. There are several reasons for this:

◆ The floor is probably not adequately reinforced with wire mesh, rebar, and girders. This may be why it's breaking up.

◆ The slab may not be adequately supported and drained on a bed of tamped gravel and sand. This, too, could be a reason why it's breaking up.

◆ Old concrete floors rarely have a moisture barrier beneath them. A barrier is important, especially if you plan to heat and cool the interior of the building after you've restored it. Absence of a moisture barrier will cause moisture to build up inside the building.

Determine whether the building is resting directly on the slab. If that's the case, you will have to raise the building as shown in Chapter 3. Break up the old slab with a jack hammer and remove it. Build or pour a new foundation and let the building down on top of it. Pour a new floor inside the perimeter of the foundation, as described on pages 97–100.

If the building doesn't rest on the slab and the foundation is in good shape, then all you need to do is remove the old slab. (*See Figure 4-12.*) Be very careful when using the jack hammer in close proximity to the foundation — you don't want to accidentally crack the foundation wall. When I've had to do this chore, I stop using the jack hammer when I'm about 1 foot away from the wall. Dig a little gravel and sand out from under the portion of the slab near the wall so the slab's hanging in air, then whack it with a sledge hammer. A chunk of the slab will break free and you can lift it out.

4-12. You can rent jack hammers from most tool rental businesses. I prefer to use the small ones, even for fairly large jobs. They don't wear you out, and they're easier to maneuver in tight quarters.

The flooring is cracked or deteriorating

Often, the problem with a floor is the flooring itself. If the flooring is in constant contact with moisture, it will rot. If it's overloaded, it will break. Mice chew their way through it to get to granaries. If the flooring has been around for any length of time, chances are that parts of it need replacement.

Replacing wood flooring — How you patch a wood floor depends on the type of flooring. You are likely to run across two different types of wood floors in barns and outbuildings: plank flooring and tongue-and-groove flooring.

Plank flooring is just that — ordinary rectangular boards fastened to joists. It rarely has an underlayment or subfloor beneath it. Barn floors and lofts are usually covered in plank flooring.

To replace a section of a plank floor, first remove the damaged boards. Pull the nails and lift the boards from the joists. (*See Figure 4-13.*) Use the old boards as templates for new ones. Measure their thickness and plane new boards to the same dimension. Rip the boards to the same width, crosscut them to the same length, then nail them in place on the joists.

What if the damage is confined to a small area and you want to replace only a portion of the boards, not the entire length? Select the boards you want to replace and score them 1 or 2 feet ahead of the damage and 1 or 2 feet behind. Draw each score mark directly over a joist. Remove the nails — be especially careful to remove them in the vicinity of the score marks. Plunge a circular saw blade into the wood and cut along the score marks. (*See Figure 4-14.*) Pry the cut portions of the boards up from the joists.

4-13. Use a nail puller to remove nails from floorboards and other constructions. Set the puller over the head of the nail and push down hard to set the claw and grab the head of the nail. Rock the puller sideways to remove the nail.

4-14. To plunge-cut flooring with a circular saw, first set the depth of cut to the thickness of the flooring. Place the saw over the cut mark, resting the saw on its toe. Retract the blade guard, turn on the saw, and slowly put it down into the wood.

Don't worry if, when you remove the cut section of the joists, you discover that you didn't "split the joist." If there isn't enough edge of the joist showing to attach new wood, fasten a short length of 2x4 to the joist that spans the opening you've created.

The procedure is a little different for tongue-and-groove flooring. Because the edges of the boards lock together, you can't simply remove the nails and pry them up. In fact, if the flooring is properly installed, you can't even get at the nails.

Begin by crosscutting the board a few feet ahead of and behind the damaged area. Score the boards over two joists and make plunge cuts with a circular saw. (*See Figure 4-15.*) Be careful not to cut through the underlayment or the subfloor if it's worth keeping.

Next, rip each board along one of the seams to cut through the tongue. Begin with a plunge cut and push the saw forward, keeping to the "waste" side of the seam (the portion that you will be removing). Pry up the first board on the waste side of the seam and pull it loose. (*See Figure 4-16.*) Continue until you have removed all the boards.

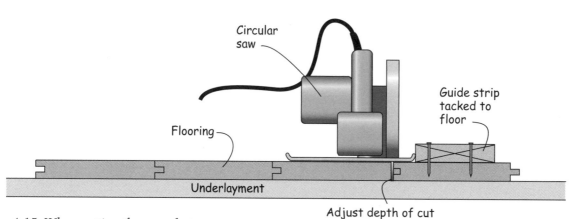

4-15. When cutting the seam between two tongue-and-groove flooring boards, keep to the waste side of the seam. It sometimes helps to temporarily nail a 1x3 to the floor to guide the saw. The board enables you to make a perfectly straight cut.

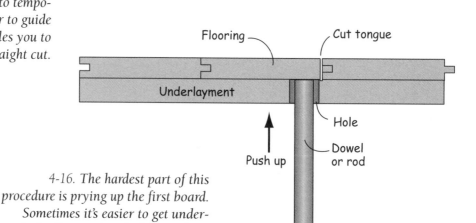

4-16. The hardest part of this procedure is prying up the first board. Sometimes it's easier to get underneath the floor and push the board up.

If you have removed the flooring and the underlayment, replace the underlayment with new boards. If necessary, attach cleats to the joists and the cut edges of the remaining underlayment to create ledges on which the new boards will rest. Plane new lumber to the exact thickness of the old and lay it in place. You can also use plywood or oriented-strand board (OSB) if it's the right thickness. Nail the underlayment to the joists. *(See Figure 4-17.)*

In some cases, you may have to create your own flooring boards if a particular style or size is no longer available. If you don't have the equipment to make tongue-and-groove joints, have a millwork company make them for you. If you can't find someone nearby who does millwork, call up a local building contractor and ask who he or she relies on for custom woodwork.

In a pinch, you can cut flooring boards without tongue-and-groove joints. The flooring will be slightly weaker in that section, but as long as you're not expecting the floor to support anything especially heavy, it should hold.

When replacing tongue-and-groove flooring, slide the first board in place, mating the tongue on one board to the groove on another. Place a scrap against the edge of the replacement board and whack it with a hammer to properly seat the tongue in the groove. Nail the board to the underlayment or the joists, driving the nails through the tongue. Repeat until you get to the last board. To get the last board in place, rip off either the tongue or the bottom of the groove, depending on how the boards are oriented. *(See Figure 4-19.)*

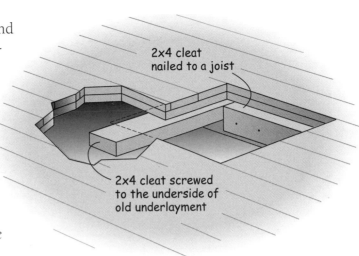

2x4 cleat nailed to a joist

2x4 cleat screwed to the underside of old underlayment

4-17. If you must replace the underlayment, create a ledge all around the opening to support it — don't just lay it across the joists. The patch will be much stronger if you tie the new underlayment to the old with cleats.

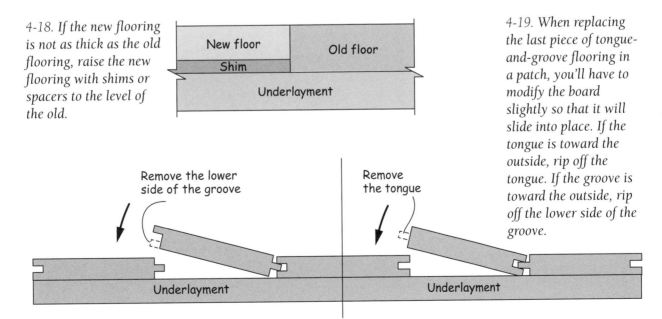

4-18. If the new flooring is not as thick as the old flooring, raise the new flooring with shims or spacers to the level of the old.

New floor

Shim

Old floor

Underlayment

Remove the lower side of the groove

Remove the tongue

Underlayment

Underlayment

4-19. When replacing the last piece of tongue-and-groove flooring in a patch, you'll have to modify the board slightly so that it will slide into place. If the tongue is toward the outside, rip off the tongue. If the groove is toward the outside, rip off the lower side of the groove.

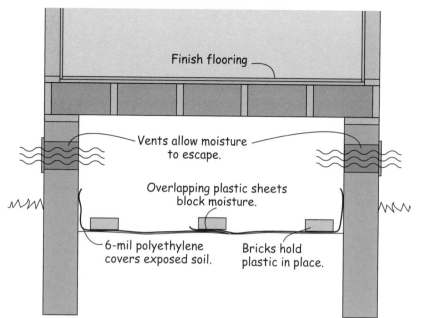

Finish flooring

Vents allow moisture to escape.

Overlapping plastic sheets block moisture.

6-mil polyethylene covers exposed soil.

Bricks hold plastic in place.

4-20. To properly ventilate the space under a floor, install vents on two opposing walls so that air can flow across the space. If this space is a crawl space with a dirt floor, cover the dirt with a 6-mil plastic tarp to hold down the moisture.

Ventilating a floor —

Often, poor ventilation is the cause of a deteriorating wood floor. Lack of ventilation allows moisture to build up in the crawl space or basement of the building. The wooden framing members and joists soak up this extra moisture, creating the right conditions for bacteria and insects to attack the wood. Consequently, the wood begins to rot.

There are several ways to ventilate the space under a floor. If you have a solid masonry or stone foundation, install vents in the sides. Choose spots along two opposing foundation walls, 10 to 12 feet apart, and carve openings with a cold chisel and a hammer. A standard vent is 6 inches by 12 inches, but check before you begin to make the openings, since sizes vary.

Remove enough material to install the vent with ¼ to ½ inch extra space on all sides. It should open and close easily so that you can regulate the ventilation according to the season — more air movement for summer, less for winter.

You can also install minivents in the header joists that rest on the foundation. Drill 1-inch-diameter holes in the middle of the header joists in between each floor joist. Install the vents in the holes with a bit of adhesive caulk. The grills in the vents are spaced to keep insects out but allow airflow for adequate ventilation. *(See Figure 4-21.)*

4-21. Minivents are about the size of a quarter. To use them to ventilate a floor, simply drill holes in the header joists and insert the vents in the holes.

Replacing a portion of a concrete floor — If you have a concrete floor that is crumbling badly, it will have to be replaced. Despite advertising claims, you can't stabilize a concrete floor that is falling apart. Some companies say they can bring cracked and crazed concrete back to life by cleaning it and pouring a skim coat over the top of the bad concrete. In my experience, this rarely works. In short order the new concrete cracks along the lines of the bad concrete underneath. If you compare the cost of the skim coat to the cost of excavating the old floor and pouring a new one, you will probably find that you aren't saving very much. Might as well do the job right.

However, if you have a small section of bad concrete, that's a different story. A bad section can be replaced as long as it can be isolated and cut off from the rest of a solid floor.

Mark out the bad section in chalk, drawing straight lines around it to create a rectangle. Rent a "cutoff saw" — a masonry saw that looks and operates much like a chain saw. *(See Figure 4-22.)* Depending on how much work you have to do and how much wet muck you can tolerate, you may want to rent a wet saw. This is the same as a cutoff saw, but it has an attachment for a hose, which reduces dust but leaves a mess in its wake. Cut through the floor along your chalklines. At the tool rental store, trade the saw for a jack hammer and break up the base section. Remove the debris. If your floor is more than 4 inches thick, you will want to drill holes and epoxy lengths of rebar in them to help join the new section of floor to the old. *(See Figure 4-23.)* Pour the floor section in the void you have created. Level it with the old float and smooth the surface.

4-22. Rent a cutoff saw to isolate and slice away a bad section in a concrete slab. Wear goggles, hearing protectors, and a respirator. As you cut, let the weight of the saw do most of the work.

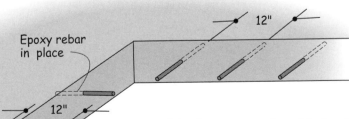

Epoxy rebar in place

12"

12"

4-23. If the floor is more than 4 inches thick, or if you want it to support a heavy load, you must tie the old floor and the new patch together with rebar. Cut lengths of about 12 inches. With a masonry bit, drill 6-inch-deep holes in the cut edges of the slab, spacing them every 12 inches. Epoxy the bars in the holes.

Patching concrete floors — If you need to patch a small hole in a concrete slab, purchase a special concrete patching mix, available at most home centers. It's slightly harder and adheres to cured concrete better than regular concrete.

Chisel away the edge of the hole to remove all the crumbling concrete and expose a clean, hard surface. Carefully clean out the hole and vacuum up all the dust. (*See Figure 4-24.*)

Mix up enough patching material to fill the hole. Wet the sides of the hole with water to help the patch bond to the concrete. Fill the hole, screed off the excess, and smooth the patch level with the surrounding concrete surface. (*See Figure 4-25.*)

Resurfacing concrete — Sometimes the surface of a concrete slab begins to deteriorate even though the slab itself is sound and solid. This often happens when standing water around the concrete freezes and thaws repeatedly. You can resurface a slab with a special resurfacing mix (similar to patching mix), although the new surface will not be as durable as the original.

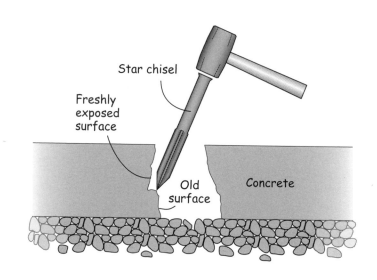

4-24. For the patch to stick to the old concrete, the sides must be as clean as possible. Use a star chisel and a hammer to expose fresh, clean, solid surface around the perimeter of the hole.

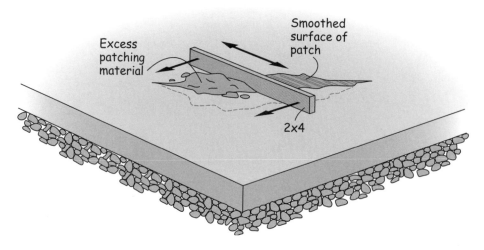

4-25. Pack the hole with patching material and screed off the excess. Rest a length of 2X4 on the floor, then make a sawing motion as you draw the board across the area. Afterward, you can smooth the patch with a trowel.

Use a star chisel and a hammer to clear away any crumbling concrete and get the slab down to solid material. If you're resurfacing a large area, rent a concrete cutoff saw and cut several long 1-inch-deep grooves, spaced every 6 inches, to help the new surface bond with the slab.

Mix up the resurfacing compound and spread it on the damaged area. Screed off the excess, then float the new surface to smooth it.

The floor must support additional weight

Often when you restore a barn or other outbuilding, you must reinforce the floor to support more weight. The runway floor in a century-old bank barn was probably built to support horse-drawn wagons and farm equipment. If the farmer was well-to-do, he might have had a steam tractor. But on the whole, his equipment was a good deal lighter than farming equipment is today. Even if you're just restoring the barn to house a small tractor or an automobile, you may be asking the floor to support more weight than it was originally designed to hold. When this is the case, you must reinforce or strengthen the floor.

Strengthening a wood floor — There are several ways to beef up a wood floor to hold more weight. The simplest is to add one or more crossbeams perpendicular to the joists, just as if you were shoring up a sagging floor. *(See pages 72 and 73.)* Support the crossbeams with floor jacks. For extra support, use steel I-beams or a laminated plywood beam instead of a solid wood beam. *(See Figure 4-26.)*

You can also install bridges between the floor joists. *(See page 73.)* These stiffen the floor and help distribute the load over a wider area.

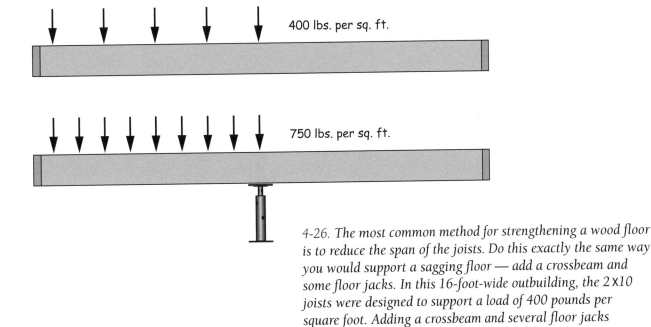

400 lbs. per sq. ft.

750 lbs. per sq. ft.

4-26. *The most common method for strengthening a wood floor is to reduce the span of the joists. Do this exactly the same way you would support a sagging floor — add a crossbeam and some floor jacks. In this 16-foot-wide outbuilding, the 2 x 10 joists were designed to support a load of 400 pounds per square foot. Adding a crossbeam and several floor jacks reduces the span to 8 feet, and the maximum load rises to 750 pounds per square foot, almost double.*

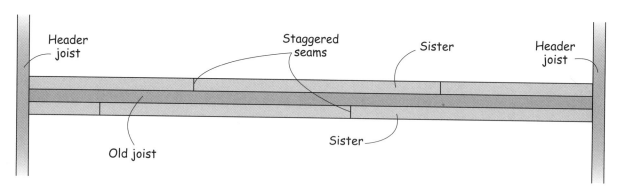

Header joist

Staggered seams

Sister

Header joist

Old joist

Sister

4-27. You can strengthen a floor by making the joists thicker. Attach sisters to each side of a joist, sandwiching the old wood between new. Since you'll probably install these sisters in sections, stagger the seams so they don't line up. If the seams are aligned, the sisters won't add much strength at all.

If you don't have room for a crossbeam or you don't want to clutter the space beneath the floor with floor jacks and posts, sister the joists. Sandwich the old lumber between new wood to increase the amount of weight the floor will support. The method is similar to sistering a joist to span a damaged or rotted section, but there are some important differences.

You must sister the entire length of the joists, not just a small area, and you must apply sisters to both sides of the joist.

The sisters must be attached in sections. You'll probably find plenty of obstacles if you try to install a sister joist all in one piece. It will be much easier to cut the sister into several lengths. When you do this, make sure the seams between the sections of one sister don't line up with the seams of the sister on the opposite side of the joist. For maximum strength, the seams must be staggered. (See Figure 4-27.)

Strengthening a concrete floor — The only way to strengthen a concrete floor is to make it thicker. A 4-inch-thick concrete slab is sufficient for most purposes, but if you plan to park large trucks and heavy machinery on your floor, you should increase the thickness to 6 inches.

The trouble is, you can't do this by simply pouring 2 inches of concrete over the 4-inch-layer that is already there — concrete doesn't work that way. You have to excavate the old floor, removing it completely, then pour a new floor of the required thickness.

Installing a new floor

If you can't repair or beef up the existing floor to suit your needs, you must replace it completely. This is a big job, to be sure, but you have several options, especially if you're installing a wood floor. You can simply take up the old flooring, leaving the underlayment in place, then install new flooring. You might be able to get by with installing the new flooring directly on the old stuff. Then again, depending on the extent of your restoration, the intended purpose of the restored building, and the extent of the deterioration, you may have to take the floor all the way back to the frame.

I've written this section to take you through all the steps, from the frame to the finish. If you don't need to go back as far as the frame or the underlayment, skip ahead and jump in at your specific starting point.

Framing a wood floor — To replace the frame of a wood floor, remove the old joists and brides. If the header joists — the boards around the perimeter of the building — are damaged or rotted, you will have to raise the building in order to replace them. (*See pages 60 to 61.*)

Cut new joists the same thickness and width as the ones you removed. To insert the new joists between the old header joists, you'll find it easier to cut each joist in two sections. (*See Figure 4-28.*) Space the joists 16 inches on center and install bridges between them.

Laying a wood floor over concrete — To lay a wood strip floor over a concrete slab, first spread a 6-mil plastic sheet over the slab to serve as a vapor barrier. Turn pressure-treated 2x4s on their sides. Using cement nails, fasten them to the concrete 16 inches apart on center. (*See Figure 4-29.*) These boards, called "sleepers," do the same job as floor joists. Nail the subfloor and floor to the sleepers.

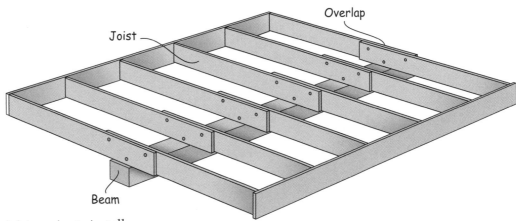

4-28. *To make the joists easier to install, cut each in two sections. I like to install a beam down the center of the floor, then install the joist sections so that the mating ends overlap over the beam. This method makes for a floor frame that's almost twice as sturdy as the original.*

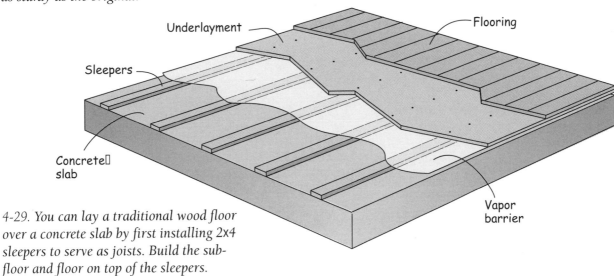

4-29. *You can lay a traditional wood floor over a concrete slab by first installing 2x4 sleepers to serve as joists. Build the subfloor and floor on top of the sleepers.*

4-30. Solid-wood tongue-and-groove flooring can be attached directly to the subfloor if you prepare the subfloor properly. Sand it with a drum sander and 36-grit sandpaper to eliminate any high or low spots, then cover it with builder's paper to provide a moisture barrier.

Preparing the subfloor — Just as a building requires a good foundation, a wood floor requires a good subfloor. How you prepare the subfloor depends on the type of floor you lay.

There are two common types of wood flooring — solid-wood tongue-and-groove flooring, which comes in long strips approximately ¾ inch thick; and hardwood plywood flooring, available in ¼- to ½-inch-thick strips. Tongue-and-groove flooring is strong enough to be applied directly over the subfloor. (See Figure 4-30.) Most plywood flooring, however, requires underlayment — large sheets of plywood or particle board fastened to the subfloor. (See Figure 4-31.) Otherwise, the subfloor cannot properly support the thin materials, so they will wear quickly. In addition, irregularities in the subfloor will "telegraph" through the floor, ruining its appearance.

Installing tongue-and-groove flooring — Before you can lay flooring, you must have at least one *baseline* — a reference line on the floor to help keep the flooring strips aligned. Snap a chalkline parallel to one wall on the builder's paper or the underlayment. (See Figure 4-32.) Measure from these baselines to check the position of the floor as you lay it.

Finish carpenters sometimes install strips of flooring around the perimeter of a room, parallel to the wall. These *borders* help frame the floor. Install the borders first, scribing and cutting the outside edges to fit the strips to the walls. (See Figure 4-33.) Leave a small gap (¼ to ⅜ inch wide) between the outside edge of the border and the wall to allow for the expansion and contraction of the floor.

4-31. Plywood flooring often requires underlayment — sheets of plywood or particleboard laid over the subfloor to smooth out irregularities and provide additional support. To install underlayment, first cover the subfloor with builder's paper. Cut the underlayment and lay it in place, leaving ⅛-inch-wide gaps between the sheets and ⅜-inch-wide gaps between the underlayment and the walls. These gaps will allow for expansion and contraction. Fasten the sheets to the subfloor with nails or screws, spacing the fasteners no more than 6 inches apart. If you use nails, drive them at slight angles, reversing the angle with each nail to prevent the underlayment from loosening.

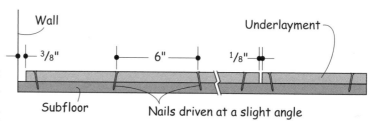

Wall · ⅜" · 6" · ⅛" · Underlayment · Subfloor · Nails driven at a slight angle

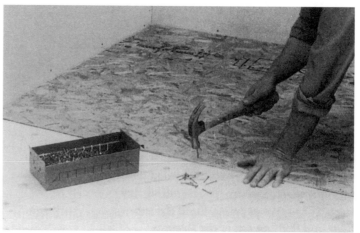

The flooring material should be kept in the building for a few days at 70 degrees Fahrenheit to allow the material to acclimate to the moisture content in the building. This step will keep your flooring material from warping or cracking after the installation.

4-32. *Before laying the floor, decide which wall you want the strips to parallel. Measure out partway into the room and put down a long, straight chalkline parallel to the wall. (You may want to snap several lines, depending on the size and shape of the room.) Use this baseline as a reference when aligning the floorboards. Don't use the wall as a reference, as walls are rarely straight.*

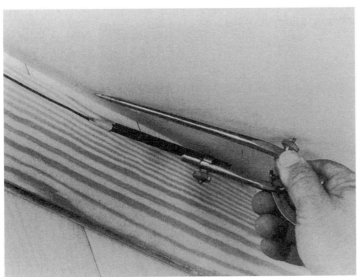

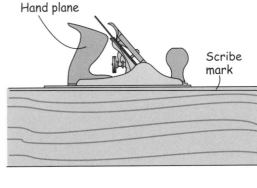

Hand plane

Scribe mark

4-33. *If your flooring plan calls for a border, install flooring strips around the perimeter of the room. Along the wall where you will begin laying the floor, install the border with the tongue facing in. Install the other borders with the grooves facing in. Leave a ¼- to ⅜-inch-wide gap between the border and the wall to allow for expansion and to compensate for irregularities in the wall. If the wall is more than ⅜ inch out of line, you will have to cut the border to fit it. Lay the border against the wall and sight down the strips to make sure they're straight. Adjust a compass so that the distance between the point and the scribe is equal to the largest gap between the wall and the border. Trace the irregularities of the wall with the compass point, scribing the border strips as you do so. Saw or plane the strips to the scribe marks.*

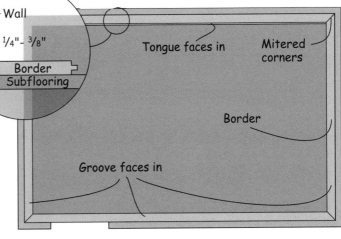

Wall

¼"- ⅜"

Border
Subflooring

Tongue faces in

Mitered corners

Border

Groove faces in

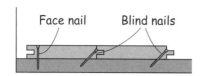

Face nail   Blind nails

4-34. There are two ways to fasten tongue-and-groove flooring to the subfloor. You can face nail (1), driving the nails through the faces of the flooring strips and setting the heads; or you can blind nail (2), driving the nails through the tongue at an angle so that they will be hidden from view when all the flooring is installed. Generally, you can blind nail most of the flooring. You only need to face nail any borders, the first course, and the last few courses. Note: If you have a lot of flooring to lay, you can rent both face nailers and blind nailers from most tool rental services.

Lay the first course of flooring parallel to a baseline, tongues facing in. Make sure the first course is absolutely straight. Nail through the face of the flooring, close to the outside edge, and countersink or *set* the heads of the nails — this is called *face nailing*. Also, nail through the tongues at an angle — this is known as *blind nailing* because the heads of these nails will be hidden when you lay the next course. (See Figure 4-34.) If you haven't installed a border, remember to leave an expansion gap between the wall and the flooring strips.

Lay the next course, blind nailing only. Offset the ends of the strips at least 3 inches from those on the first course and tap them into place with a rubber mallet. Lay the remaining courses by using a *beater block* to seat the boards before you nail them. (See Figure 4-35.) To save time, you may want to start *racking* — laying out several courses at once, then nailing them in place. (See Figure 4-36 and 4-37.)

4-35. As you lay each successive course of flooring, you must seat the tongues and the grooves together as tightly as they will go. To do this, place a scrap of flooring (called a beater block) against the strip and whack it with a mallet. Slide the beater block sideways a foot or so and repeat until you have seated the entire course. Sight down the course to make sure it's straight and measure from the closest baseline to check that it's parallel. Finally, blind nail it in place.

4-36. The job goes faster and you get better results if you rack several courses at a time before you install them. Select strips from the bundle of flooring and arrange them on the subfloor in the order in which you will install them. Match the grain and the wood tones to form pleasing patterns.

Sight down each course as you beat it into place to check for straightness. Also, measure its position relative to a nearby baseline. If it's not parallel to the baseline, gradually correct the error over several courses. You can also taper flooring strips, if necessary. *(See Figure 4-38.)*

When butting flooring strips to borders, install *splines* between the boards. *(See Figure 4-39.)* Also use splines to reverse direction of the tongues when you must back lay flooring into a closet or an alcove. *(See Figure 4-40.)*

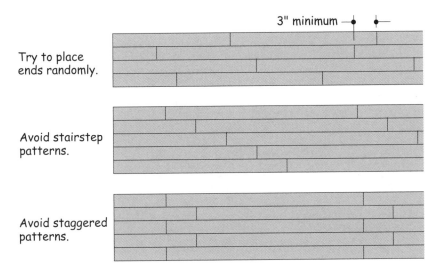

Try to place ends randomly.

Avoid stairstep patterns.

Avoid staggered patterns.

3" minimum

4-37. For strength, the ends of the strips in each course should be offset at least 3 inches from the ends in the previous course. Unless you want to create a precise geometric pattern in the floor, place the ends randomly. Imprecise and partial patterns will be distracting. A random look is easy to achieve when you rack the courses before you install them; it's much more difficult if you lay out and nail only one course at a time.

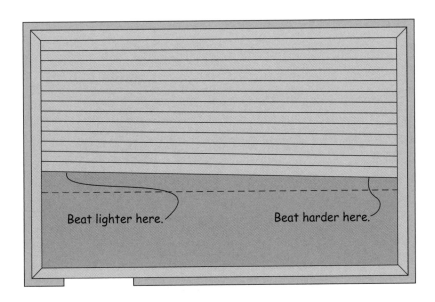

Beat lighter here.

Beat harder here.

4-38. If you discover that the flooring is misaligned with a baseline, you can correct the problem over several courses. When seating the flooring, beat the strips a little harder where they are too close to the baseline and not quite so hard where they are too far away. If the misalignment is extreme or if you reach the other side of the room and find the walls are not parallel, saw a taper in the last course.

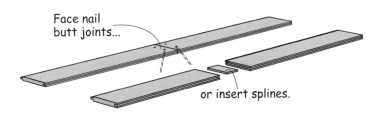

Face nail butt joints...

or insert splines.

4-39. When you butt flooring strips end to end or against borders, install splines between the parts. When you can't use splines, face nail the ends of the boards where they butt together.

As you approach the end of the job, face nail the last few courses. Because you'll likely be up against a wall, you won't have room to use a beater block. Instead, push or pull the strips into position with a wood scrap. (*See Figure 4-41.*)

4-40. Occasionally, you must back lay flooring in alcoves and closets, reversing the direction in which you were working when you laid the main portion of the floor. To do this, you must also reverse the direction of the tongues and grooves. Install a spline in the flooring groove where you will start back laying. Use this like a tongue, fitting the groove on the next course to the spline.

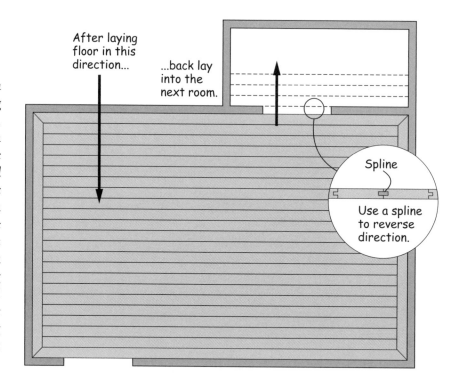

After laying floor in this direction...

...back lay into the next room.

Spline

Use a spline to reverse direction.

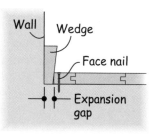

Wall Wedge

Face nail

Expansion gap

4-41. You won't be able to seat the last few flooring courses with a beater block because the wall will be in the way. Instead, seat them by using scraps as a lever and a fulcrum. Place one or more short scraps (the fulcrum) against the wall, and insert the end of a longer scrap (the lever) between the wall and the last course. Push the lever toward the wall so that it pivots on the fulcrum, forcing the flooring strips into place. When the space between the wall and the flooring grows too narrow to let you use this method, rip the bottom of the groove off the last course so you can slip it in place. Wedge it against the previous course and face nail it.

Laying plywood flooring — Like solid-wood flooring, hardwood plywood flooring also fits together with tongues and grooves. For that reason, it's installed in much the same way. Lay down baselines and align the plywood strips parallel with them, fitting the tongues and grooves together. Stagger the ends of each course, and try not to let obvious patterns develop.

Plywood strips are thinner than solid-wood flooring, so the tongues are more fragile. You cannot beat them in place; even a rubber mallet may prove too aggressive when you're installing thin flooring. Instead, you must slide the strips together with fingertip pressure only.

You cannot blind nail thin strips because, once again, the tongues are too fragile. Plywood flooring is designed to be laid with flooring adhesive, or *mastic*. Spread mastic with a grooved trowel, applying enough for several courses at a time. *(See Figure 4-42.)* Face nail the first course so it won't shift. For successive courses, press the flooring into the adhesive until it "grabs," then slide the tongues and grooves together. *(See Figure 4-43.)*

Wait until the mastic sets completely before you walk on the floor or sand it. While the adhesive is still soft, the flooring can easily scoot out of alignment. Depending on the type of adhesive, it may take a day or more to harden.

4-42. *Most plywood flooring is installed with mastic, a special flooring adhesive. Spread a thick coat on the floor with a grooved trowel and wait a few minutes for the adhesive to flash — it will glaze over and feel slightly rubbery. However, don't let it dry out. The adhesive must grab the flooring when you press it in place. If you can't feel it gripping, it's too dry, and you must scrape up the mastic and apply new.*

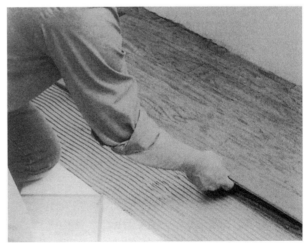

4-43. *Lay the courses of plywood as you would tongue-and-groove flooring. Rack several courses and apply just enough mastic to the floor to lay them. Slide the strips into place, then repeat. As you work, check the position of each course relative to a baseline. If a course is not aligned properly, shift the flooring to bring it back into alignment. The mastic does not set quickly; it will remain soft enough for you to move the flooring for many hours.*

Installing a new floor over an old one — In some
cases, you may want to install a new floor over an old
one. I've covered a few old plank floors with OSB just to
give them more strength. I've also put down "floating"
floors on concrete slabs in outbuildings that were being
restored as workshops. A floating floor is made of plywood tongue-and-groove strips. The strips are covered
with a thick hardwood veneer so that they can be
sanded and finished like ordinary hardwood flooring. To
install this flooring material, simply glue the tongues in the grooves and
let the strips float on the concrete like a plywood carpet. If you have a
workshop with a concrete floor that's tiresome to stand on, a floating floor
is a great way to get some relief. It's also a good way to convert an outbuilding for living space — the hardwood floor has a much more inviting
look than a concrete slab. *(See Figure 4-44.)*

*4-44. A floating floor isn't
attached to the supported material below it; it simply rests on top
of it. Often, the flooring strips are
cushioned underneath, making
them more comfortable to stand
on without sacrificing strength.*

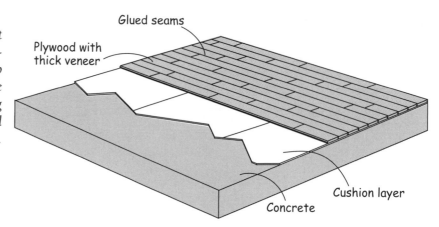

You can also install a traditional tongue-and-groove floor over an old
plank floor. If the planks are in good shape, you can use them for underlayment — just tack down builder's paper to the planks before you install
the flooring to prevent drafts and create a moisture barrier. If planks are
cupped and worn, install plywood or OSB underlayment over the planks,
then install the flooring on top of the new underlayment. *(See Figure 4-45.)*

*4-45. You can install a new
tongue-and-groove floor over an
old plank floor using the old floor
as underlayment if the planks are
in relatively good condition. You
will have to either replace damaged planks or cover them with
sheet material. If they are in
extremely bad shape, it may be
better to strip the planks off and
put down new underlayment.*

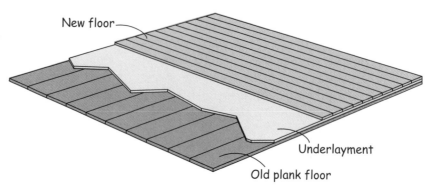

Finishing floors

Nowadays, much commercial flooring comes prefinished; all you have to do is install it. However, there are still many types of floors that must be finished after the floor- ing is laid. There are four major steps to finishing a floor: rough sanding the surface, filling the gaps between the boards, final sanding, and applying the finish.

1 Wait until you're ready to apply a finish before sanding the floor. Don't let it sit for several weeks after sanding it, since it will get dirty. Rough sand the surface first, starting with 50-grit sandpaper and working up to 100-grit. Sand the main body of the floor with a drum sander, traveling with the grain. (You can rent these sanders from most tool rental services.) Sand small areas and edges with an edge sander or random orbit sander. *Note:* Before you sand, mask off entrances to other rooms and put a fan near a window to exhaust air to keep the dust from creeping into other parts of the house.

2 Clean up the dust from the rough sanding and inspect the floor for gaps between the boards. Fill these gaps with a filler made especially for floors. There are two types, premixed fillers made to match particular woods or stains and those you mix yourself with wood dust from the rough sanding. Apply the filler with a putty knife or a trowel, forcing it into the gaps. *Note:* Gaps are a common problem with all floors, but especially with parquet. The parquet sections are not perfectly square, so there are small gaps where they butt together.

(continued on next page)

Finishing floors — *continued*

3 After the filler dries, sand the floor a final time with 120-grit paper. (If you're finishing a parquet floor, work up to 180-grit.) Don't be afraid to do a little hand sanding when necessary — machine sanders won't reach every nook and cranny. When the floor is uniformly smooth, thoroughly clean and vacuum the surface. Wipe the floor with a tack rag to remove all the sanding dust.

4 Ventilate the room so that finish fumes won't collect in the house. Apply oil-based finishes with a bristle brush, brushing with the grain to spread the finish out in a thin, even coat. Apply water-based finishes conservatively with a foam pad or short-nap applicator. (If you flood a raw wood floor with a water-based finish, it will swell and distort.) Let the first coat dry to the touch, give it a light sanding with 240-grit or 320-grit sandpaper, then apply the second coat. ***Note:*** Carefully read the instructions that come with the finish.

Pouring a new concrete floor — Excavate the old concrete or the dirt floor you will be replacing. Be sure to excavate about 6 inches for a 4-inch-thick floor, 8 inches for a 6-inch-thick floor and so on. You want to leave room for a sand "backfill" that should be at least an inch thick. If you are pouring a slab — a combined floor and foundation — you must excavate trenches around the perimeter of the floor. See page 66 for more details.

Build a form around the perimeter of the pour, using ¾-inch plywood and 2x lumber. *(See Figure 4-46.)* The top of the form must be even with the desired elevation of the floor — use a level to check this. In some cases, you may want the floor to slope slightly toward a floor drain or an opening. Garage floors, for example, commonly slope toward the doors. Brace the forms every 2 feet to keep the poured cement from pushing them outwards. If you are pouring up to an existing wall, you won't need to build a form for that section. But you do want to snap chalklines to indicate the floor elevation.

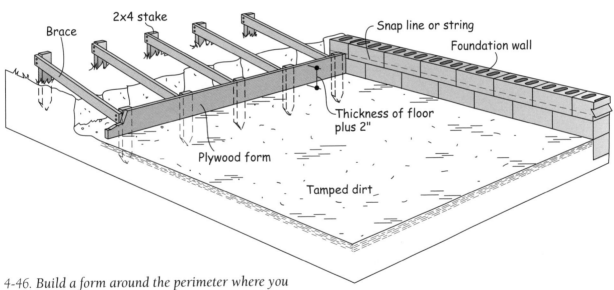

Brace

2x4 stake

Snap line or string

Foundation wall

Thickness of floor plus 2"

Plywood form

Tamped dirt

4-46. Build a form around the perimeter where you will pour the floor and brace it. If the floor will support the structure, you must also excavate and build forms for footers.

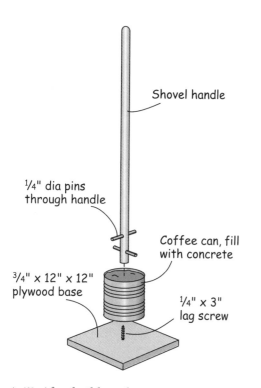

Shovel handle

¼" dia pins
through handle

Coffee can, fill
with concrete

¾" x 12" x 12"
plywood base

¼" x 3"
lag screw

4-47. After building the forms, and before you backfill, tamp down the dirt to compact it as much as possible. This will prevent the floor from settling unevenly and cracking overmuch.

Tamp the excavation with a vibrating tamper or a hand tamp. *(See Figure 4-47.)* The more you tamp the better. Backfill the area with sand 1 to 2 inches deep. Level the sand with a screed board (a 2x4 long enough to ride on the tops of your form boards). Set the screed board aside — you'll need it later for screeding the concrete.

Install plastic sheeting at least 6 mils thick over the backfill. This serves as a vapor barrier between the ground below and your workspace above. It also keeps radon gas from seeping up through the ground into your workspace. If you are pouring next to an existing floor foundation wall, now is the time to install rebar anchors as shown on page 83.

Lay out reinforcing wire or fabric mesh in the bottom of the pour. (Whether you use wire or the less expensive fabric mesh depends on your local codes — check them.) Support the mesh on a few inches about the vapor barrier. You can purchase special holders to do this, but I find stones or bricks work just as well. The mesh should end up in the middle of your concrete pad when it's poured. *(See Figure 4-48.)*

Make careful measurement of your completed forms. Using these, a concrete supplier will help you determine what kind of mix you will need and how much. Order the concrete. If the truck cannot drive within a few yards of the pour, or if you don't want a heavy truck driving across your lawn, hire a concrete pump and operator to pump the concrete to the

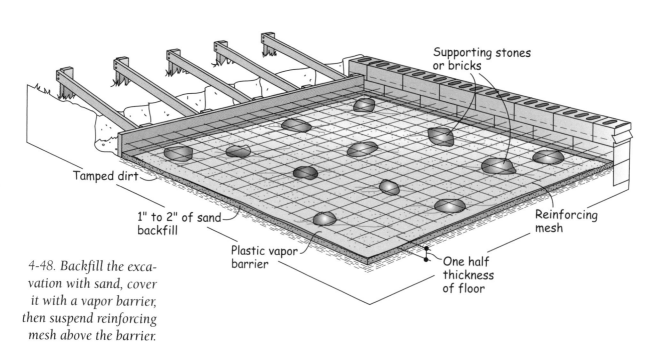

Supporting stones
or bricks

Tamped dirt

1" to 2" of sand
backfill

Plastic vapor
barrier

Reinforcing
mesh

One half
thickness
of floor

4-48. Backfill the excavation with sand, cover it with a vapor barrier, then suspend reinforcing mesh above the barrier.

pour. This will cost about $100 for a half day. The drawback is that the mix must be more liquid than you would normally use for this type of pour. If yours is a small pour, you can lean on a few friends to form a wheelbarrow brigade to carry the cement.

With your forms in place and your screed boards ready, begin the pour at the furthest point from the concrete truck. Using your eye judge how much you need and spread it out accordingly. At 6- to 8-foot intervals stop and screed the concrete. (*See Figure 4-49.*) The screed height is really the finish height. Work quickly, but don't get in a hurry. The concrete takes over an hour to begin to set up.

After the rough screeding is done, use a bulldog float to smooth the cement. (*See Figure 4-50.*) A float forces the gravel in the mix beneath the surface and causes the "cream" (the soupy stuff) to float to the top. Go over the surface of the pour just once — don't keep floating it.

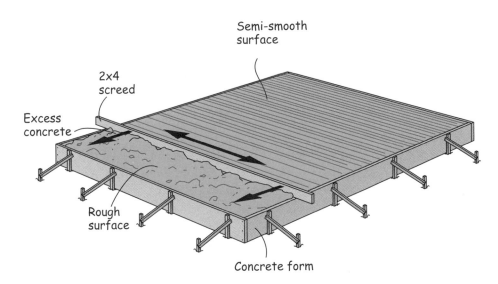

Semi-smooth surface

2x4 screed

Excess concrete

Rough surface

Concrete form

4-49. As you pour, scrape or screed the wet cement to level the pour with the forms or elevation lines. It helps to slide the screed board back and forth as you draw it across the cement.

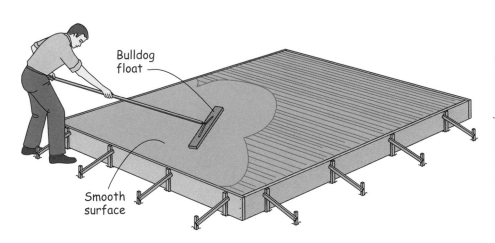

Bulldog float

Smooth surface

4-50. Before the concrete sets, smooth the surface with a bulldog float — this is a wide trowel on a long handle. You can rent them from most tool rental businesses.

After two or three hours, the concrete will be firm enough to support your weight if you stand on a wide board or a piece of plywood. If you want a super-smooth surface, place a board on the concrete floor and smooth it with a hand float or a power trowel. *(See Figure 4-51.)*

Let the slab set up for 24 hours, then cut control joints across the surface every 4 to 8 feet using a cut-off saw and a masonry blade. *(See Figure 4-52.)* These joints help to control where the concrete will crack. A concrete floor or slab will crack; this is a fact of life. Control joints keep the cracks straight and make them less noticeable.

4-51. Once the cement is hard enough to support your weight while standing on large board, hand float the surface, smoothing it to the desired finish. Work backwards across the floor, always smoothing the area where you last placed the board.

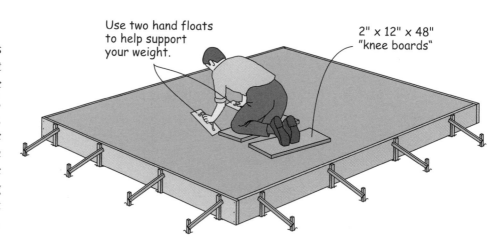

Use two hand floats to help support your weight.

2" x 12" x 48" "knee boards"

4-52. Cut control joints in the slab after the concrete has set up for 24 hours. With a cut-off saw and a masonry blade, cut grooves about ¾ inches deep. Guide the saw along a 2x4 to keep the grooves straight.

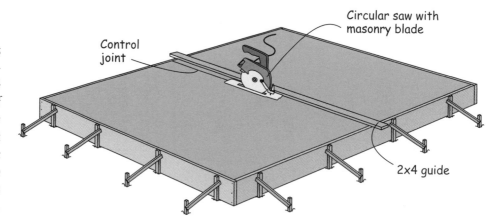

Circular saw with masonry blade

Control joint

2x4 guide

Structure and framing

Most barns and outbuildings have an interior skeleton, a wooden frame that holds up the walls and roof. There are exceptions, of course. Old buildings with walls made from stone or masonry often have no wall frames. Neither do log structures, unless you want to consider the interlocking logs as a kind of framework. But a typical

5-1. Older barns often have unconventional and possibly unreliable structures.

wooden building has either a post-and-beam frame or one made from dimensional lumber. And, not surprisingly, the frame members sometimes get tired, saggy, weak, and rotten with age. When this happens, you must either reinforce the frame or replace the impaired pieces.

Fixing structural problems

Before we get started, let me warn you that replacing and repairing structural members is not simply a matter of swapping out the worn-out parts. Every frame member supports a certain amount of weight, and when you adjust or remove a member you often have to temporarily support that weight with something else. If you don't, the building may be damaged. It could even collapse. The rule of thumb is to build a temporary framework to take the weight while you're replacing the permanent member.

Gussets and cleats

Of course, not every structural repair requires you to replace a frame member. Often, weak or damaged members can be reinforced with gussets and splices. A gusset is a thin piece of wood or metal that spans a

5-2. A broken mortise-and-tenon joint in a post-and-beam structure.

joint or seam that has come apart. A cleat is a board that reinforces a cracked or broken frame member.

For example, if you have a post-and-beam structure and one of the tenons has broken off in a mortise (*See Figure 5-2*), you can fix the break without remaking and replacing the tenoned member. Draw the parts back together with a well-placed jack or come-along, then join them by nailing metal gussets on opposing sides of the joint. (*See Figure 5-3.*) In some cases, you may have to remove some siding boards to reach both sides of the joint.

If a beam is cracked in the middle, cut two cleats from boards that are roughly the same width as the broken part. Brace the beam so it's straight, then nail the cleats to opposing sides of the beam. (*See Figure 5-4.*) For maximum strength, position the cleats to best resist any racking forces on the beam.

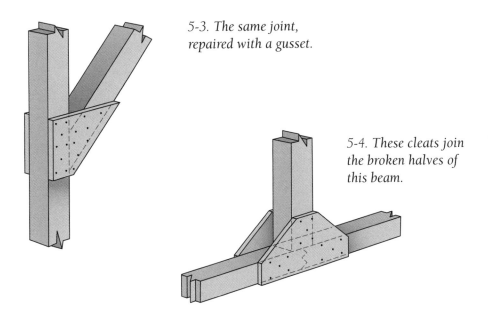

5-3. The same joint, repaired with a gusset.

5-4. These cleats join the broken halves of this beam.

E X T R A   H E L P

When attaching a wooden gusset or cleat, stagger the nails so they aren't in a row. This adds strength and helps prevent the repair part from splitting.

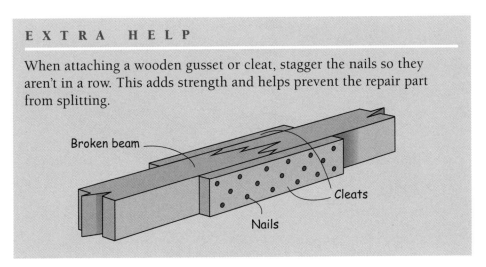

Broken beam

Cleats

Nails

Splices

You can't always patch up a broken joint or frame member with a gusset or a cleat. Occasionally you must splice in new wood to replace the damaged portion, or you may even have to replace a missing portion of a board. When this is the case, you have several choices (*See Figure 5-5*):

◆ Cut away the damaged portion of the wood and make a new piece that fits the gap. The ends should all butt together. Fasten the new wood to the old with gussets and nails.

◆ If you don't have room for or don't want to apply gussets, make scarf joints. Cut long, matching bevels in the ends of both the old wood and the new, fit them together bevel to bevel, and secure the pieces to each other with bolts or lag screws. Typically, a scarf joint has a slope of 1 to 12 (1 inch rise for every 12 inches of run), but I have cut bevels with slopes of 1 to 18 when I needed extra strength.

◆ If you don't need the strength that a scarf joint affords, make simple end laps by cutting interlocking steps in the ends of the boards. Bolt the lap joint together.

The structure is leaning

My grandfather used to tell a joke about my great-grandfather's hog barn. He would say that anytime great-granddad wanted to know the wind direction, he'd just look out the window to see which way the barn was leaning. There was nothing wrong with the barn structurally. But

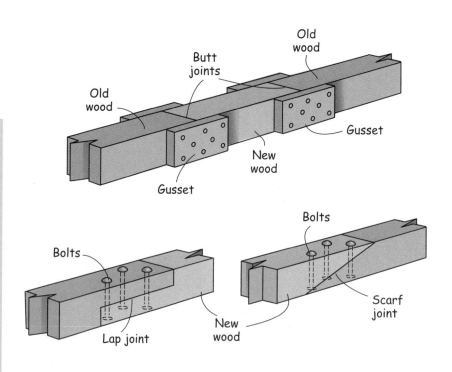

5-5. *Three methods of splicing boards and beams.*

over the years, wood shrinks and swells with the seasons and nails slowly rust away. The fasteners became loose in their holes, and the structure began to lean.

Putting a leaning structure back to rights looks like a big task, but it's actually a fairly straightforward structural repair. To begin, note the direction of the lean and the angle the barn walls make with the horizon. One wall will be acute, the other obtuse. Nail a long 2x10 near the foundation of the obtuse wall. Drive the nails into the framework, not just the siding. Let about 6 inches of the 2x10 protrude past the corners in both directions. Do the same for the acute (opposite) wall, but nail the 2x10 just below the eaves. Install

5-6. *A barn with a serious, but repairable, lean.*

four large eyebolts, one in each of the protruding ends of the 2x10s. This will give you a handy purchase to attach a pair of come-alongs. String the cables from the come-alongs diagonally from the low to the high eyebolts. (*See Figure 5-7.*)

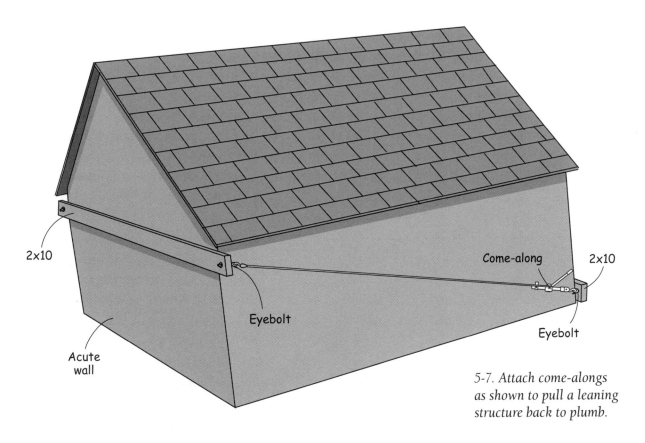

2x10

Come-along   2x10

Eyebolt

Eyebolt

Acute wall

5-7. *Attach come-alongs as shown to pull a leaning structure back to plumb.*

5-8. Use diagonal braces to keep the structure plumb once you've pulled it back to rights.

Put a little tension on the come-alongs, tightening them until you see the structure just begin to move, or until you start to hear it creak a little. Ratchet the come-alongs in turn, first one and then the other. It may be a good idea to round up some help for this task; it will wear you out running from one side of a big barn to the other.

Once the come-alongs are holding the structure to prevent it from leaning any further, remove the diagonal bracework from the walls along which the come-alongs and cables are stretched. *Don't* remove all the bracework, just the diagonal members from those two walls closest to the come-alongs. You might be able to get by with just loosening the diagonals. In many cases, there may not be any diagonal members to worry about. Many carriage barns and smaller outbuildings were built using the horizontal wooden siding to brace the structure. If you're putting an old post-and-beam structure back to rights, just tap the pegs loose that hold them in place.

With the bracework removed, begin to tighten the come-alongs again, one at a time, little by little. They will start to pull the leaning walls upright. As this happens, check the corners with a level or a plumb bob. When all the corners are straight up and down, give your helper a pat on the back. Tell him he can breathe now.

Once the structure is standing straight, replace (or add) diagonal bracework. If you don't intend to finish the walls on the inside of the barn, you can simply nail long 2x4s at an angle across the studs or vertical supports. Slope each 2x4 so one end is high in a corner and the other reaches to the sill board. (*See Figure 5-8.*) If you plan on finishing the walls later, you'll have to "cut in" the diagonals, making lap joints wherever they cross the studs. (*See Figure 5-9.*)

Once all of the bracework is in place, remove the come-alongs and the 2x10s.

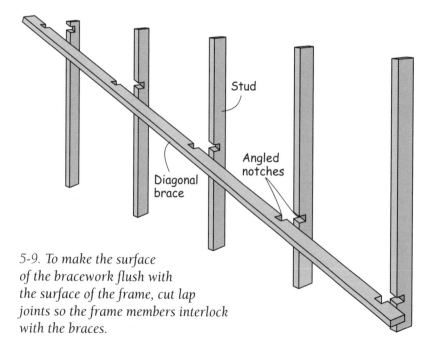

Stud

Angled notches

Diagonal brace

5-9. To make the surface of the bracework flush with the surface of the frame, cut lap joints so the frame members interlock with the braces.

EXTRA HELP

If you're righting a particularly large barn or a structure that seems particularly unstable, use four 2x10s and four come-alongs instead of just two. Nail 2x10s high and low on each of the angled walls, then attach the come-alongs diagonally so each pair makes a big "X." Tighten all four come-alongs to take the slack out of the cables. Then, as you pull the structure upright, loosen one set of come-alongs as you tighten the other set.

A bracework primer

The most common way to brace a rectangular structure such as a typical barn or an outbuilding is to *triangulate* the framework and diagonal members, breaking the rectangles up into triangles. Rectangles will rack. Gravity and other forces pull a large rectangular frame into a parallelogram, distorting the structure or making it lean. Triangles, however, are rigid. They can't be pulled out of shape. Use a diagonal to divide a single rectangle into two triangles, and the rectangle becomes rigid. Below are several suggestions for bracing different types of structures.

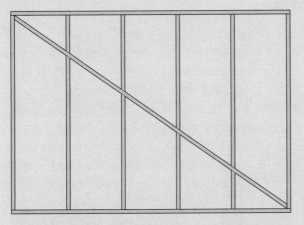

Commonly, a wall is braced with a single long diagonal that is either cut into the vertical supports or applied over them.

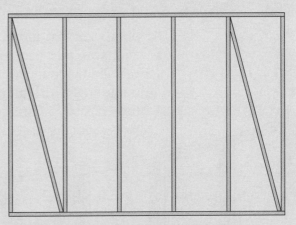

Sometimes only the corners need bracing. Run long diagonals from the corner post to the nearest vertical support member.

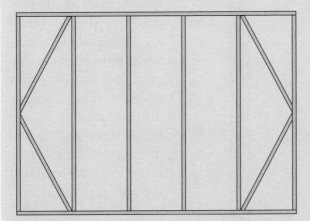

Alternatively, brace the corners with short diagonals running from the middle of the corner post to the top and bottom of the nearest vertical members.

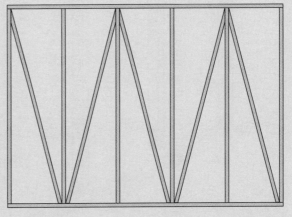

If you need exceptional strength in a supporting wall, use "W" braces — diagonals that run up and down between the vertical members.

Sagging walls

The walls of old buildings often sag somewhere along their length. The foundation may have settled unevenly, or the sill and vertical supports may have rotted away in one area from insect damage or contact with the ground. I once saw a barn where the owner had piled manure in a corner for as long as he had owned the farm. The manure eventually ate away the corner post, and the corner of the barn sagged badly.

To fix a sagging wall, first determine the cause. If the problem is an uneven foundation or an inadequate foundation that allows a portion of the barn to rest on the ground, then you have a two-part repair on your hands: you need to repair both the wall and the foundation. If the foundation is sound and the sag is due to rot, breakage, and insect damage, then the repair is simpler. For those repairs requiring that you fix the foundation, turn to Chapter 3. I cover just the structural portion of the repairs in this section.

Determine where the wall sags and how far the sag has spread. To do this, measure down from the top plate to a point about eye level on one vertical support — let's say this distance is 10 feet. Put a chalk mark at that point. Do the same for the remaining vertical support members in the vicinity of the sag (and a little beyond it), measuring down exactly 10 feet from the top plate and making chalk marks. Look down the length of the wall, sighting along the chalk marks. Where the wall sags, the mark will dip. The marks will show you the exact shape and extent of the sag. (See Figure 5-10.)

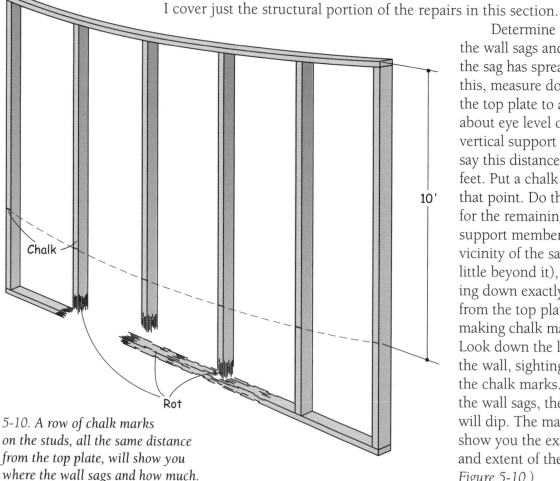

5-10. A row of chalk marks on the studs, all the same distance from the top plate, will show you where the wall sags and how much.

Remove the siding in the vicinity of the sag, starting at the bottom and continuing up about 2 feet or until you are well past any rot. Nail a short (2-foot long) 2x10 crosspiece to each vertical support about 2 feet above the ground. Place two concrete blocks near the base of each vertical member, one inside the building and the other outside. Place jacks on top of the blocks and adjust them so the tops of the jacks just touch the bottom edges of the crosspieces. *(See Figure 5-11.)* You can purchase foundation jacks for this task, or you can borrow hydraulic jacks and scissor jacks from every friend and relative within driving distance.

Raise the jacks a little at a time, beginning where the wall sags the most. As you do so, continually refer to the chalk marks you made on the vertical members. When the lowest chalk mark is even with the two on either side, begin raising all three. When those are even with the flanking pair, raise five. And so on, until all the chalk marks are even. *(See Figure 5-12.)* At this point, if you need to repair or replace the foundation, do so.

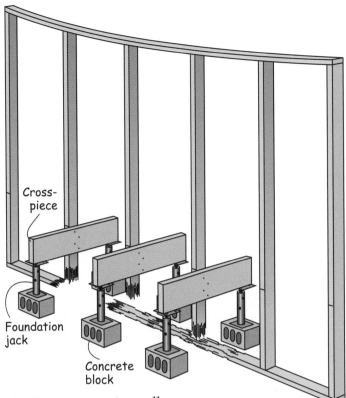

5-11. *To raise a sagging wall, attach crosspieces to the frame members, then place jacks under the ends of the crosspieces.*

E X T R A   H E L P

To check that the chalk marks are dead even and level, stretch a string between the two outermost marks and hang a string level from it. All the marks should align with the string, and the bubble should be centered in the level.

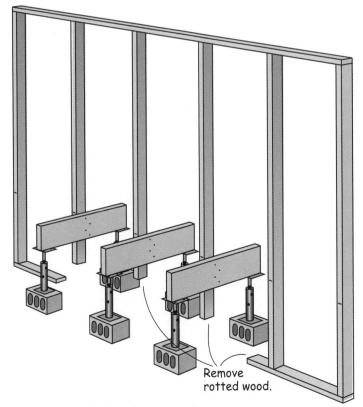

5-12. *Jack up the low frame members until the chalk marks are all in a straight, level line.*

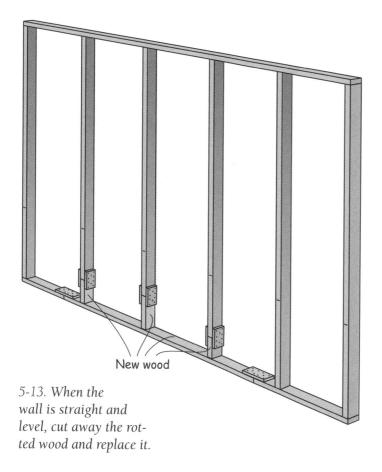

5-13. *When the wall is straight and level, cut away the rotted wood and replace it.*

New wood

5-14. *Built-up dirt and debris can cause a sill to rot to the point where it must be replaced.*

Cut away the rotten portion of the sill (if there is a sill) and the vertical members. Splice a new sill board to the old one, if needed, then splice in short lengths of beams or dimensional lumber to replace the rotten portions of the vertical members. Toenail the spliced sections of the vertical parts to the sill. (*See Figure 5-13.*)

Let the wall down slowly, lowering the jacks *in turn* a little at a time. Once the wall is sitting solidly on the foundation and you're reasonably happy with the facelift, remove the jack and the crosspieces. Consider adding some diagonal braces in the vicinity of the former sag to help strengthen the wall. The structure will be a little weaker where you've been moving it around, and braces will help prevent any problems that might arise because of that. After adding the braces, replace the siding.

Rotted sill

Occasionally you'll see an outbuilding where the structure is sound enough that the walls haven't sagged, but the sill is rotting away nonetheless. You must replace the sill, of course, but the first step is to find what's causing it to rot. Most likely, the wood is in contact with the ground. Even though the building may be sitting on an adequate foundation, sometimes the debris that accumulates outside — leaves, branches, dirt that blows in or washes up against the structure — raises the ground level over the years. If this is the case, you must clear away the accumulated dirt and debris to bring the ground back to its original elevation. (*See Figure 5-14.*)

Once you're sure that the foundation is adequate and the sill is not in contact with the ground, you must replace the rotted portions. Follow the same procedure for fixing sagging walls. Jack up one side of the structure 1 or 2 inches to lift the sill off the foundation, remove the deteriorated wood, replace it with new wood, and let the barn back down gently.

I once repaired a garage in a small town. The structure itself was built at the beginning of the 20th century, but it sat on a stone foundation that was laid in 1820. Over the succeeding years as the town grew, the land around the garage rose. Construction debris and runoff from a nearby parking lot raised the ground until it was well above the sill. I dug away the dirt that had accumulated in a century and a half only to find I had created *another* problem. Essentially, I had dug a trench around the garage. When it rained, the water ran off the parking lot and flooded the trench. This, in turn, thoroughly soaked the sill and the floor of the garage. To completely solve the problem, I had to dig down further, then lay drainage tile and a bed of gravel to absorb the runoff. (*See Figure 5-15.*)

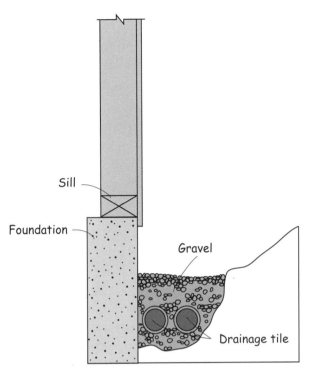

5-15. *If the landscape rises around the structure, install a drainage system so the sills will remain dry.*

Rotted sill log

Replacing the sill is a common repair in log cabins. These structures were rarely placed on a proper foundation when they were first built. Most were simply set on cornerstones — one big stone in each corner. Many were erected right on the ground. Even if they were moved to proper foundations after they were first built, chances are some rot has begun in the sill logs. And once it's begun, it will eat away the wood like a cancer.

5-16. *Log cabins rarely had full foundations, so rotting logs near the ground is a likely problem.*

Replacing a sill log is similar to replacing a sill board. However, because of the way a cabin is constructed, it requires a little more ingenuity. You can't simply strip away the siding to expose the frame, then fasten a few crosspieces in place to jack up the structure. Instead, you must fasten long 2x8s vertically to the inside and outside of the cabin, running up at least three or four courses of logs. Use 6-inch-long lag screws to secure these vertical members. Nail a crosspiece and a sturdy brace to each vertical member near its bottom end. Place concrete blocks and jacks under each crosspiece, then raise a wall of the cabin a little bit at a time until you can pull the sill log free. (*See Figure 5-17.*)

Carefully copy the old sill log, making a new one. Typically, sill logs were made from oak, even though the rest of the cabin may be hewn from poplar or another wood. Settlers knew that oak sills were more resistant to rot and insects. Measure the diameter of the log needed, then purchase a similar log from a sawmill. If the cabin is very large, you may have to splice two or more logs together to make a sill of the proper length. Hew the log to the same shape as the original sill and cut the ends to copy the joinery.

Now comes the fun part. Gather a few of your friends to help move the sill log into place. For a 20-foot-long oak log, you'll need eight helpers with shoulders of Atlas and the patience of Job; I know that for a fact. Place loops of rope beneath the log and station the helpers two to a rope loop, one on each side of the log. In unison, lift the log and walk it into place.

The easy part is getting it to the cabin. The hard part, and the one that will take you and your friends most of a day, comes next. Scoot the log between the jacks and concrete blocks. Once the log is in position, scoot it out again when you find you've reversed the ends. Turn the log around, scoot it back in, then scoot it out a second time when you discover it's 2 inches too long. Recut one of the joints, then scoot it back in position for a third and final time.

Sometime during this procedure, one of your friends will suggest that a great deal of effort could have been saved by measuring the

EXTRA HELP

The distance between the concrete blocks and the outside of the cabin must be slightly more than the width of the sill log, otherwise you won't be able to scoot it out.

5-17. *To lift a log wall, fasten huge brackets made from 2×8s to it and place jacks under the brackets.*

Logs

2x8 bracket

Jack

Concrete block

Rotted sill

One log width

sill log while it was still in position under the cabin, then clearly marking the inside and outside surfaces of the replacement.

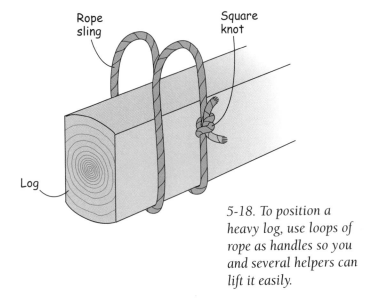

5-18. To position a heavy log, use loops of rope as handles so you and several helpers can lift it easily.

CONSIDER AN ALTERNATIVE

You don't have to replace the rotted sill log with another, if you'd rather not. The folks who restored this beautiful crib barn simply replaced the rotted log with a stone foundation. If necessary, they could have continued that foundation up one or two courses, creating walls that are part stone and part log.

Hewing logs and beams

To replace a log in a log cabin or a beam in a post-and-beam structure, you sometimes have to hew a log to make a copy of the part. The logs for cabins are commonly hewn on two sides to make them the same width; beams are hewn on all four sides to make them square. Hewing is a time-consuming chore, but not a difficult one.

1 Stake the log to prevent it from rolling. If the log is tapered, prop up the tapered end so the center or *axis* of the log is approximately level. Fasten crosspieces to the stakes on either side of the log. The crosspieces must be level, even with each other and slightly below the top of the log. To determine how far below, measure the diameter at the largest end and divide by six. The top edges of the crosspieces should be ⅙ of the log diameter below the top of the log.

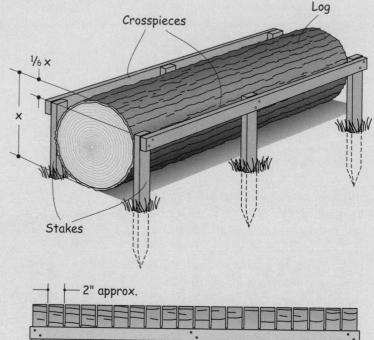

Crosspieces

Log

⅙ x

x

Stakes

2 With a chain saw, score the top of the log every 2 inches, carefully cutting down to the top edge of the crosspieces.

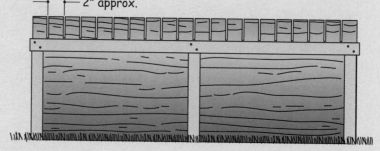

2" approx.

3 Using an adze, a broad axe, or a side axe, chop away the wood where you have scored the log. To remove any evidence of having used a chainsaw, chop a little deeper than the score lines. This will leave the surface looking as if it were hand-hewn.

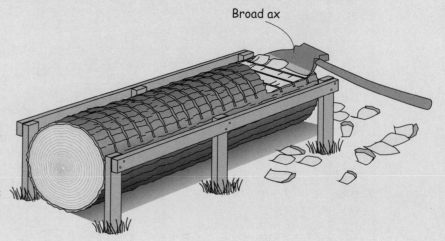

Broad ax

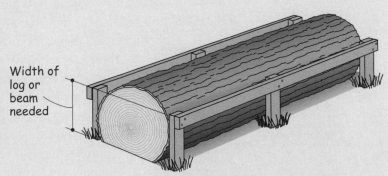

Width of log or beam needed

4 **Alternative:** If you aren't concerned about the appearance of the surface, you can use a chainsaw and an "Alaskan sawmill." This is a fixture that fastens to the bar of your chainsaw and lets you mill beams and boards. Guide the saw mill fixture along the top surface of the crosspieces.

5 Roll the log over so the flat side is resting on the ground. Drive the stakes down and arrange the crosspieces so the distance from the flat side of the log to the top of the crosspieces is equal to the width of the log or beam you want to replace. Repeat the process, flattening the opposite side.

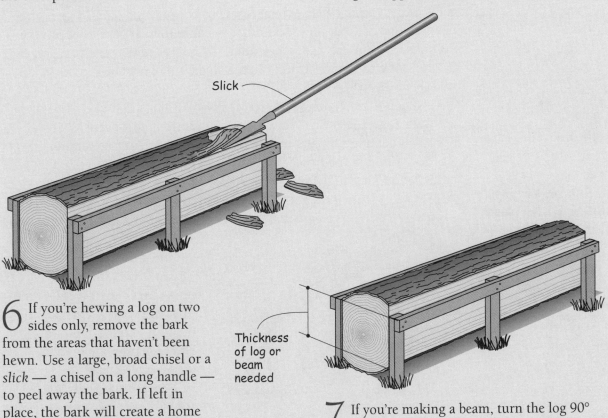

Slick

Thickness of log or beam needed

6 If you're hewing a log on two sides only, remove the bark from the areas that haven't been hewn. Use a large, broad chisel or a *slick* — a chisel on a long handle — to peel away the bark. If left in place, the bark will create a home for insects, and these will eventually eat away the log.

7 If you're making a beam, turn the log 90° and move the stake in to keep it from rolling onto a flat side. Position the crosspieces and flatten the third side. Roll the log 180° and repeat for the last side, hewing the log square.

Bowed walls

The framework of an older structure is not always tied together as soundly as builders tie a barn frame together nowadays. This is especially true of the roof frame. Instead of making rigid trusses, old-time carpenters raised a ridgepole running the length of the barn, then leaned the

rafters against the ridgepole. Over the years, the weight of the roof pushes out on the top plate, causing the walls to bow outward. At the same time, the roof "bellies," sagging in the middle.

Like sagging walls, this is a two-part problem. To solve it, you have to fix both the walls and the roof. For some ideas on how to reinforce a roof with insufficient trussing, see Chapter 8. To fix the bowed wall, simply pull it back to rights.

Install an eyebolt in the top plate of the bowed wall, toward the middle of the wall, or wherever the bow is most pronounced. Install a second eyebolt in the opposing wall, directly across from the first. *(See Figure 5-19.)* If the structure is extremely large, you may need two or more eyebolts in each wall. When I straightened the walls of the carriage barn where I now have my wood shop, a civil engineer advised me that there shouldn't be more than 16 feet between the eyebolts, or between the eyebolts and the corner of the structure. My 30-foot-long carriage barn required a single set of eyebolts, right in the middle of the walls.

Attach a come-along between the two eyebolts, then begin to draw the walls together *slowly*. "Slowly" is an important part of these instructions. Unlike other repairs in this chapter, this particular operation affects a huge portion of the frame, moving studs, sills, rafters, and many other parts. If you have to move each wall more than 3 or 4 inches, it would be a good idea for you to do so over several hours.

5-19. To prepare to pull two bowed walls plumb, install eyebolts in the top sills.

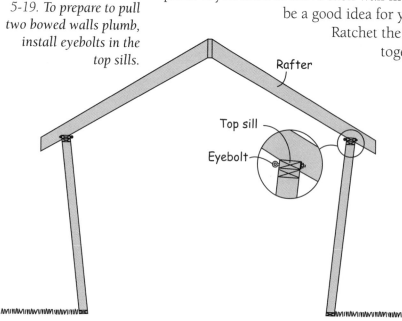

Rafter

Top sill

Eyebolt

Ratchet the come-along so you draw the walls together just 2 or 3 inches or until you hear the frame start to "groan," then stop and give the frame members 20 or 30 minutes for the new tensions in the frame to reach equilibrium. Draw them in a few more inches and repeat.

When the walls are straight, you must brace them so they don't bow out again. You could install braces on the outside, running from the top plate to a place on the ground a few feet out from the building to create a buttress. This, in fact, is the purpose behind flying buttresses on old cathedrals. I've also seen buttresses like these on old aircraft hangers and some barns, mainly in the south. But the more common solution is simply to tie the eyebolts together with a cable or a chain. (*See Figure 5-20.*) You may also want to reinforce the roof by tying the rafters together beneath the ridgepole.

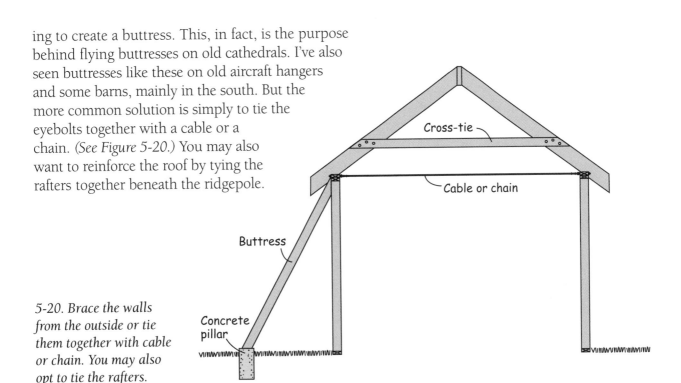

5-20. Brace the walls from the outside or tie them together with cable or chain. You may also opt to tie the rafters.

Cable tools

The quickest and least expensive way to tie two eyebolts together is to use multi-strand steel aircraft cable. This is available in sizes from ¹⁄₁₆ inch up to ½ inch or larger in diameter. Use cable ¼ inch in diameter for small structures such as garages; use ⅜- to ½-inch-diameter cables for larger buildings.

Tie off the ends of the cable by looping them through the eyebolts, then securing them with cable clamps. For safety, use *two* clamps to fasten each cable end. For a more permanent tie-off, use cable ferrules. Slide these metal collars over the end of the cable *before* you loop it through the eyebolt. Double the cable back on itself, slide the ferrules over the cable end, and crimp them in place with a swaging tool.

Adding to or modifying a structure

In addition to repairs and restorations, I've had to modify a structure when the purpose of the building changed. One of the most common modifications is the removal of load-bearing posts or walls to open up the space inside a garage or barn. I've also had to cut quite a few holes in outside walls for new doors and windows. And occasionally, when the space is inadequate, I've had to enlarge a structure or add a room.

Removing a post

Before you remove framing on the inside of a structure, first determine if the frame is *load-bearing*. Does it support any structural weight or is it important to structural integrity? If you have any doubts about this, consult an architect or ask your county engineer. You'll need someone with engineering skills looking over your shoulder on this job, anyway. It's best to consult an expert sooner rather than later.

If the frame members are load-bearing, decide how you're going to support the load or maintain structural integrity once they are removed. Chances are, you will have to brace or beef up the existing structure to redistribute the load. The most common way to do this is to add a beam or truss with sufficient strength to handle the weight in the absence of the load-bearing frame members that will be removed. (*See Figure 5-21.*) Be sure to ask your architect or engineer just how much movement you should tolerate when the load is shifted to the new beam or truss.

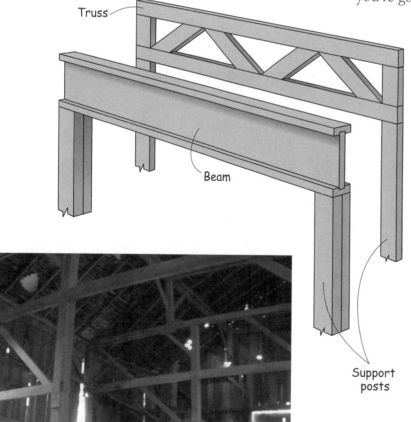

Truss

Beam

Support posts

5-21. A beam or a truss can support the weight of a structure once a portion of a load-bearing wall is removed.

I once removed several posts that were supporting a hay loft in a small barn. The barn had been converted to a workshop, and the owner was tired of dancing around the posts — he wanted an unobstructed work space. So we installed a 4-inch-thick, 12-inch-wide laminated plywood beam to span the 22-foot-long space under the loft. This beam supported the weight that had been resting on the posts.

Here's how we did it. First, we hoisted the beam in place next to the smaller wooden joist that ran across the posts. Then we secured the joist to the beam with lag screws every 12 inches. At the ends of the beam, we installed 4x4 support posts, set back into the outside walls. (See Figure 5-22.)

With the beam and supports in place, we removed the old posts. This is the exciting part, because as each post comes out, the weight of the structures transfers to the new beam and you discover just how good your architect or your county engineer really is. To hedge my bets, I took out the posts in stages. First, I installed a long floor jack next to each post, pressing up against the joist. (See Figure 5-23.) I made a single cut through the post, removing about ⅛ inch of stock with a reciprocat-

ing saw. Then I slowly retracted the floor jack. As the weight of the barn loft shifted to the new beam, the saw kerf in the post closed a little, but I could still see daylight where I had cut. This told me the beam was adequate to support the weight, and it had been installed correctly. Only then did I completely remove the post.

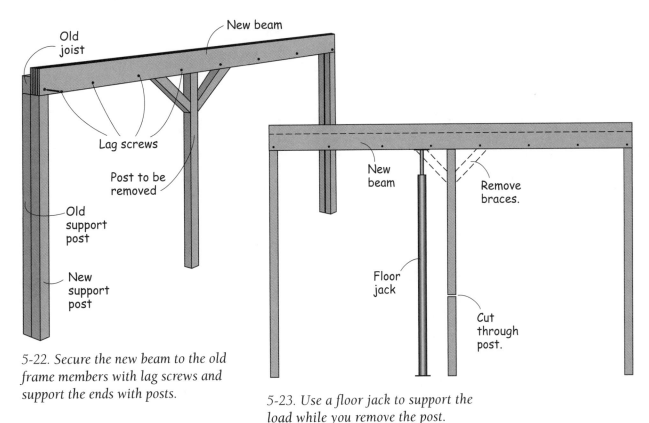

5-22. *Secure the new beam to the old frame members with lag screws and support the ends with posts.*

5-23. *Use a floor jack to support the load while you remove the post.*

What if the kerf had closed completely? I would have extended the jack again and made a second cut in the post, removing about ¾ inch of wood — that was the maximum amount of movement my architect advisor told me I should tolerate. Then I would have let the jack down for a second time while watching to see if this larger gap closed. If the loft had sagged this much over a 20-foot span, I would have concluded that the beam or the installation was inadequate. At this point, all you can do is crank the floor jack up again and bring back your architect for another consultation. Usually the fix for this problem is to add a second beam, or to replace the first beam with a bigger, stronger one.

Removing a wall

Removing a supporting wall is similar to removing a post. Once again, the structure above the wall must be supported when the wall is gone. And once again, the solution to this problem is to span the gap with a beam.

Install the new support member parallel to the wall frame, and as close to it as possible. Support the ends with posts. *(See Figure 5-24.)* If you don't want these posts to be visible after you make this modification, set them back into the outside wall frames. Shore up the beam with floor jacks spaced 8 to 10 feet apart.

Carefully remove the members of the wall frame. In the middle of the span, hang a plumb bob from the new beam. Remember to ask your

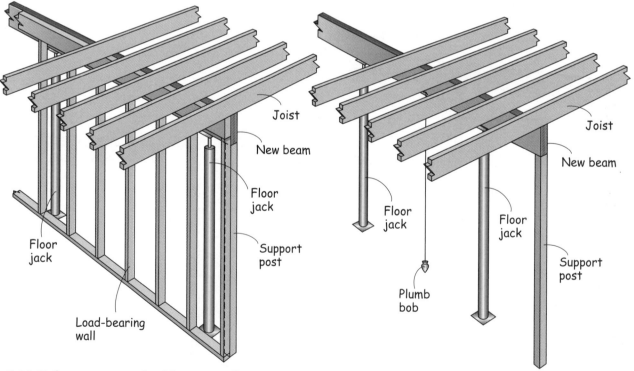

5-24. *Before removing a load-bearing wall, install the beam that will take the load once the wall is gone. Place several floor jacks along its length.*

5-25. *Slowly retract the floor jacks, transferring the load to the beam. Monitor how much the beam sags with a plumb bob.*

architect or engineer how much they would allow the beam to sag over its span. Adjust the cord so space between the tip of the plumb bob and the floor is equal to the movement you can live with. Slowly retract the jacks, shifting the weight to the beam. *(See Figure 5-25.)* Check the bob as you do this. If it touches the floor, run the jacks back up again and bring your architect back for another discussion. Just as when you are removing a post, the fix for this problem is usually to add a second beam, or to replace the first beam with a bigger, stronger one.

Cutting or enlarging an opening

To add doors or windows to a structure, you must create openings for them. When replacing doors and windows you sometimes have to enlarge the opening or move it. The method for all of these tasks is the same.

Begin by outlining the new opening on the inside of the structure, then remove any interior wall covering and insulation. If you are enlarging the existing opening, remove the old door or window, along with the jambs (the boards that frame the door or window).

Brace the structure above the planned opening to prevent it from sagging when you cut away the framework. There are three ways to do this, depending on how the building is constructed. The most common is to use a *whaler.* Cut a 2x12 a little longer than the opening you want to make and fasten it to the framework a foot or more above the top of the planned opening. Position floor jacks at each end of the whaler and extend them so they're tight against the whaler and the floor. *(See Figure 5-26.)* This will support the structure while you cut an opening and install the new framework.

5-26. Use a horizontal brace or **whaler** to temporarily support a structure when cutting an opening in a load-bearing wall.

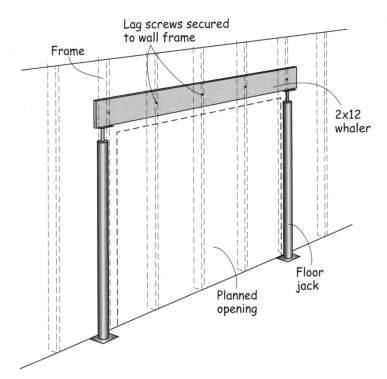

Lag screws secured to wall frame

Frame

2x12 whaler

Floor jack

Planned opening

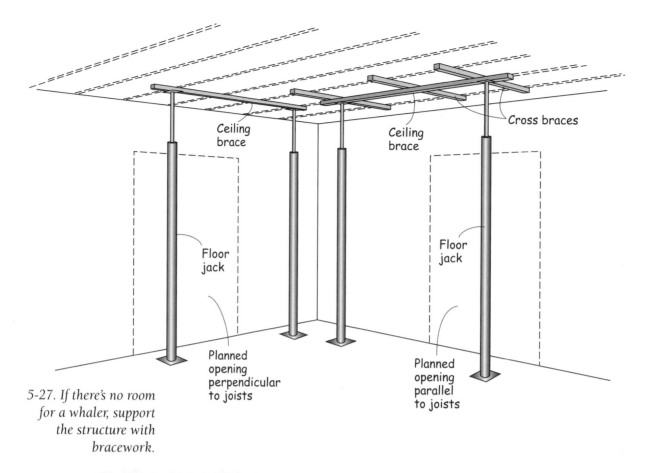

5-27. If there's no room for a whaler, support the structure with bracework.

5-28. When the structure is braced, remove the frame members one at a time.

If you don't have enough space between the top of your planned opening and the ceiling to attach a whaler, use *shoring* to support the structure while you make the opening. If the planned opening is *perpendicular* to the ceiling joists above it, cut a single 4x4 about two feet longer than the opening is wide. This serves as a ceiling brace. Hold or fasten the ceiling brace under the joists 2 to 3 feet out from the wall. Position floor jacks under the ends of the brace and extend them. (See Figure 5-27.)

If the planned opening will be parallel to the joists, make a ceiling brace and several 4x4 cross braces. Attach the cross braces to the joist, then attach the ceiling brace to the cross braces. Position the floor jacks under the ends of the ceiling brace. (See Figure 5-28.)

When the structure is properly braced, cut through the old framework with a reciprocating saw. Remove the studs or posts one at time, checking the whaler or the shoring each time.

After removing the old frame members, install the new ones. To frame a door, install king studs and jack studs on either side of the opening, then lay a header across the jack studs. Toenail the ends of the cut-off frame members to the header. If they don't reach, splice short extensions to them. Framing in a window is pretty much the same, but you must add a sill plate and cripple studs to the bottom of the framework. *(See Figure 5-29.)*

Make sure that the jack studs are plumb and that the header and sill are level so the opening is reasonably square. And it doesn't hurt to measure the opening once or twice before you drive the nails home just to make sure the door or window assembly will fit. The opening should be ¼ to ½ inch taller and wider than the door or window assembly.

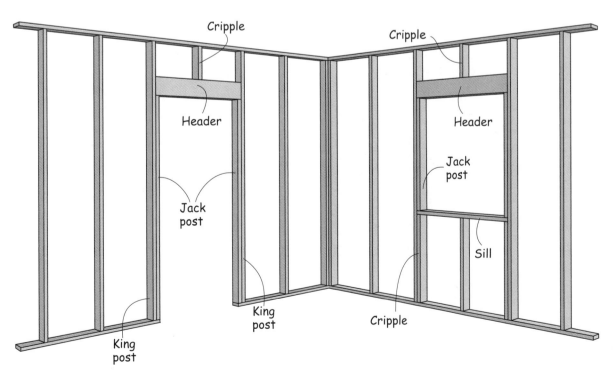

5-29. Typical framework for doors and windows.

Wait until the last possible minute to cut the hole in the exterior wall. Leave the siding in place for as long as possible to keep whatever you have stored in the barn or garage safe from the elements and secure from people who are less than honest. When you're ready to install the door or window, cut through the siding with a circular saw or a reciprocating saw. (*See Figure 5-30.*)

For a masonry wall, measure and mark the opening on both the inside and outside of the wall. Then score the masonry along the outline. (*See Figure 5-31.*) Use a circular saw with a carborundum masonry blade and cut just ¼ to ½ inch deep. Work slowly. If you feed the saw too fast, the blade will heat up and grow dull. If you feel the electric motor heating up, you're probably working the saw too hard. Take a break to let the motor cool down, then begin again, cutting more slowly.

CONSIDER AN ALTERNATIVE

To create an opening in a wall made of brick or block, first decide how wide the opening will be. You can cut openings 40 inches wide or less without installing temporary supports. If the opening is wider than 40 inches, attach a whaler to both the inside and outside of the structure. Fasten it in place with lag screws, sinking lead anchors in the masonry every foot or so.

5-30. Cut away the exterior siding from the opening just before you install the door or window.

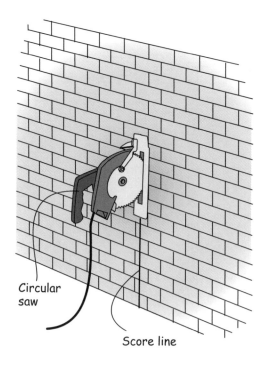

Circular saw

Score line

5-31. Score the opening in a masonry wall with a circular saw and a masonry blade.

When you have scored the open-ing, you must install a *steel lintel* at the top. This is an L-shaped steel beam that supports the masonry above the open-ing; it's available from most large retail-ers that sell building materials. You can also buy it from steel suppliers. Create a lintel channel by removing some masonry above the opening. Position the point of a cold chisel at the mortar line and strike it with a large hammer. This will pulverize the mortar. Continue until the brick or block comes loose, pull it free, and do the same for the next in line. If the wall is made of bricks, remove a double row. If it's made of concrete block, remove a single row. Chip away all the old mortar sticking to the masonry at the top of the opening, then lay the lintel in place. (*See Figure 5-32.*)

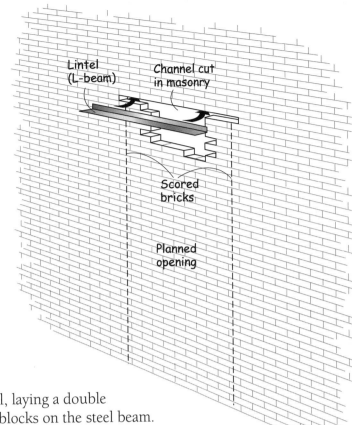

5-32. *Install an L-shaped steel beam or **lintel** in the channel you've cut in the wall.*

Replace the masonry above the lintel, laying a double row of bricks or a single row of concrete blocks on the steel beam. The sides of the beam should be no thicker than a layer of mortar, so you will be able to do this without splitting the bricks or blocks lengthwise.

When the lintel is installed, complete the opening. Remove the masonry from the top down, working out from the center toward both sides. When you need to split a brick or a block, place the point of the cold chisel in the shallow line that you scored and hit it sharply with a hammer. The brick or block will break along the line. (*See Figure 5-33.*) Don't worry if the break is a little ragged. You won't see this after you install the framework for the door or window.

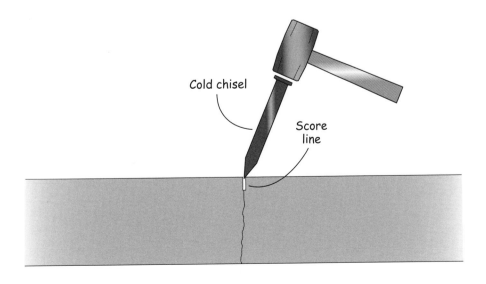

5-33. *To split a brick, place the point of a cold chisel in the score line and strike it with a hammer.*

Spans and loads

When repairing or modifying a structure, you must *span* a lot of gaps with lumber. A joist, for example, spans the distance from sill to sill. A rafter spans the gap from ridgepole to top plate. A header spans the gap from jack post to jack post. Every time you span one of these gaps, you must decide what type of material to use and what size to make it.

The answer to this question can be quite complex. What you use to span a gap depends not only on the dimension of the gap and the strength of the materials, but also on the load that these materials must support. This can vary quite a bit depending on how you intend to use the building and where you live. For example, if you're converting a small storage shed into a workshop and you intend to install heavy machinery, the floor joists must be adequate to support the additional load. If you're adding a room to a barn in a northern latitude, you must take into account the weight of the snow that is likely to accumulate on the roof when deciding on the materials for the rafters.

There are four types of loads to consider:

◆ The *dead load* is the weight of the building materials alone.

◆ The *live load* is the weight of the people who occupy the building and what is stored in it — furniture, machinery, vehicles, animals, and so on.

◆ The *snow load* is the weight of the accumulated snow and only applies to roofs.

◆ The *wind load* is the force of the wind pressure against the outside surfaces of the building, and it only applies to these surfaces.

These loads are usually calculated in pounds per square foot (psf). Here's how to estimate each load type:

Dead loads

List the different types of materials in the building once you've completed the planned repairs or changes. Consult the chart below and record the estimated pounds per square foot that these materials contribute to a structure. Then simply add them up. For example, a floor frame with 2x8 joists (3 psf) and covered with plywood sheathing (3 psf) and a hardwood floor (4 psf), must support 10 psf of dead load — 3 + 3 + 4 =10.

Dead loads	psf
Framing materials (16″ on center)	
2x4	2
2x6	2
2x8	3
2x10	3
Floor, wall, and ceiling materials (per inch thickness)	
Softwood	3
Hardwood	4
Plywood	3
Concrete	12
Stone	13
Brick	13
Drywall	5
Paneling	3
Roofing materials	
Softwood (per inch thickness)	3
Plywood (per inch thickness)	3
Asphalt shingles	3
Asphalt roll roofing	1
Asphalt (applied)	6
Wood shingles	3
Steel	2
Slate	12
Tile	19

Live loads

The live loads are specified by the building codes in your area. Ask your county engineer for a copy. Here's a fairly typical list for a residential area:

Residential live loads	psf
First floor, residence	40
Second/third floor	30
Balconies	100
Stairs	100
Garages	50
Roofs	
Slope less than 4/12	20
Slope between 4/12 and 12/12	16
Slope greater than 12/12	12

Wind load

The wind load varies greatly with your geographical location. In some areas, it may be negligible for small structures. In others, it may be enormously important. Check your local building codes.

Snow load

The snow load depends on your geographical area. Consult the map below and determine the number for the area where you live. This is the 50-year snow load — the weight of the largest recorded snow accumulation during the last 50 years. Multiply this number by 0.8 and compare it with the live loads for roofs in your area. You must use the larger of the two numbers when repairing or building a roof. For example, if you live near Dayton, Ohio, and the pitch on your barn roof is 12/12, the live load is 12 psf and the snow load is 20 psf. When you multiply 20 times 0.8, the result is 16. This is larger than 12 psf, so use 16 psf as the combined live and snow load.

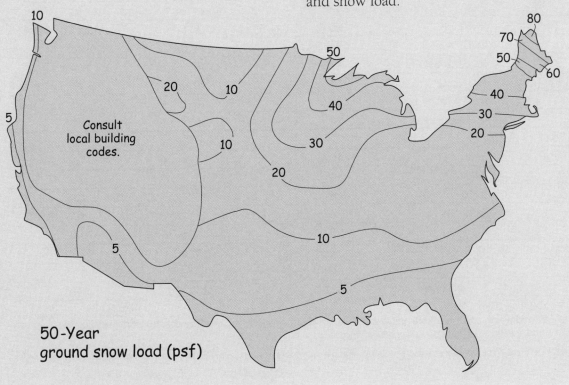

Consult local building codes.

50-Year
ground snow load (psf)

Spans and loads — *continued*

Length of span

To estimate the maximum span length for a specific material under a given load, add the dead load, live load, snow load, and wind load (if any) to get the total load on the portion of the structure that you're repairing or changing. Remember, you need only add the snow load for roofs and the wind load for outside surfaces. Once you know the total load, consult the chart below.

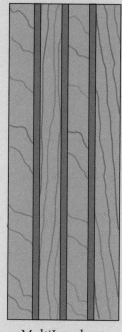

MultiLam beam

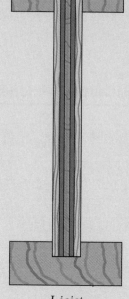

I-joist

Spacing	Material	Small load (40 psf)	Medium load (50 psf)	Large load (60 psf)
16" o.c.	2×4	8'	7'	6'
	2×6	12'	11'	10'
	2×8	16'	14'	13'
	2×10	20'	18'	17'
	2×12	24'	21'	20'
	9-1/2" I-joist*	17'	16'	14'
	12" I-joist*	23'	20'	17'
	4×12 MultiLam†	30'	26'	22'
	6" Steel beam	36'	32'	28'
24" o.c.	2×4	7'	6'	5'
	2×6	10'	9'	8'
	2×8	13'	12'	11'
	2×10	16'	15'	14'
	2×12	20'	18'	16'
	9-1/2" I-joist*	15"	13'	11'
	12" I-joist*	17'	15'	13'
	4×12 MultiLam†	24'	21"	18'
	6" Steel beam	30'	27'	24'

*Prefabricated wood I-joists are more stable and will span greater lengths than ordinary construction-grade lumber. They are available on special order through most building suppliers.

†A MultiLam is a thick beam made by laminating multiple layers of plywood. Like I-joists, these are available on special order.

Adding room to a structure

If you need more space in an outbuilding, it's a reasonably simple matter to add a room. The only trick is in tying the new room to the structure.

Decide how large you want to make the room, and lay the foundation for it outside the structure. If the room will have a concrete floor, pour a pad.

Create a door or an opening for the structure into the planned room. Follow the steps that begin on page 121 and continue right up until it's time to cut away the exterior siding. Stop short of this last step — there's no sense in exposing whatever you have stored in the old building until the new room is under roof.

Nail corner posts to the structure, one on either side of the planned door. These corner posts must be attached to the building frame, so if there isn't a stud or a post behind the siding where you want to attach the new members, you must install them. Build the wall frames out from the corner posts. (*See Figure 5-34.*)

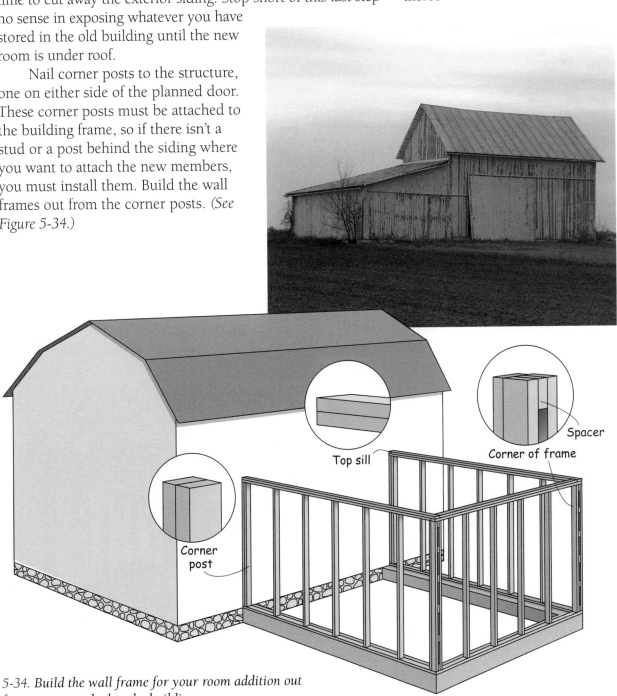

5-34. *Build the wall frame for your room addition out from posts attached to the building.*

To make a flat shed roof, attach a horizontal ledger to the old structure, spanning the width of the new room. Run the rafters from the ledger to the top plate. (*See Figure 5-35.*)

For a gabled roof, the procedure is more involved, particularly when the roofs join at right angles. You must create valleys where one roof meets the other. Build and attach the roof trusses over the new room. Run a string along the peaks of the trusses to the old roof. Also stretch strings along the bottom ends of the rafter to the eaves. This shows you where the new roof will join the old one. (*See Figure 5-36.*) If you draw lines between these points, they should form a triangle.

Wait for a stretch of good weather before this next step. When you can be reasonably certain that you have two or three days in a row with little chance of rain, snap chalk lines on the roof between the juncture points you've identified. Cut through the roofing 2 to 3 inches *outside* the chalk lines and peel up the roofing inside the triangle. Snap new chalk lines (the old ones will have disappeared with the roofing) and attach 2x4 ledgers to the roof so the outside edges are even with the lines. Run a ridgepole from the peak of the roof truss nearest the structure to where the ledgers join. Cut *crippled* (short) rafters to fit between the ridgepole and the ledgers, then nail them in place. (*See Figure 5-37.*)

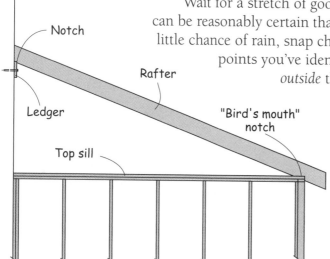

5-35. Build the roof frame out from a horizontal ledger.

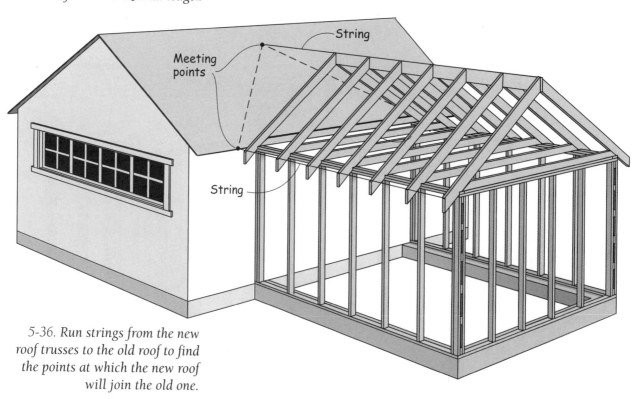

5-36. Run strings from the new roof trusses to the old roof to find the points at which the new roof will join the old one.

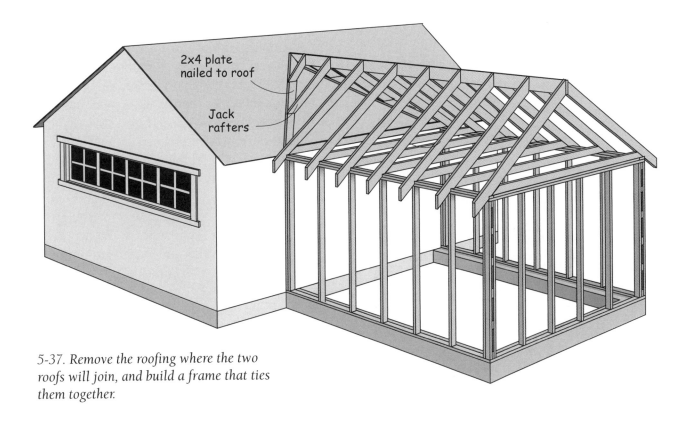

2x4 plate
nailed to roof

Jack
rafters

5-37. Remove the roofing where the two
roofs will join, and build a frame that ties
them together.

Cover the roof frame with sheathing and roofing. If there are
valleys where the new roof joins the old one, line the valley with flashing
before you attach the roofing. The completed roof will offer enough
protection from the elements that you can afford to cut away the siding,
opening up the new room to the older structure. Attach the sheathing
and siding to the walls, then install the windows and exterior doors.

If you're attaching a new frame-construction room to a masonry
structure, fasten the corner posts and ledger with lag screws and lead
anchors. However, to add a masonry room to a masonry structure, you
must tie the brick or blocks together. Every fourth course of bricks or
every second course of blocks, fasten a galvanized L-bracket to
the older masonry wall with a lag screw and a lead
anchor. The bracket should protrude out over the
new masonry, resting on the last course laid. Lay
more bricks or blocks on top of it, imbedding
the protruding arm of the bracket in the newly
laid wall. (See Figure 5-38.)

Brackets

5-38. Tie a new masonry
wall to an old one with
metal brackets imbedded
in the new work.

New
wall

Nineteenth-century log cabins were small by modern standards; consequently, those of us who own one often need to add a room or two. Adding a log room to an existing log structure can be a troublesome job. That's why the original settlers never did it. When they outgrew their cabin, they either razed it and built a new dwelling, or they added a room built with a more updated method of construction. However, I have grafted log rooms to log cabins, and the patients have survived the transplants.

Find some suitable logs from which to build the new addition. Typically, I've located smaller cabins and attached them to larger ones. But you can also tear down an old barn and use the beams, or hew new logs.

Lay a foundation for the new room. Remember that the logs in adjoining walls must be "staggered" so they will interlock. Design the foundation so each log in the addition meets the cabin *between* the logs in the adjoining wall. Remove the chinking in the area where you plan to attach the new room. Cut large slot mortises between the old logs, 4 to 8 inches high and as wide as the logs in the new addition. (*See Figure 5-39.*) Temporarily put the sill logs in place and mark where the ends of the logs meet the mortises. Cut tenons to fit the mortises and slide the logs in place. (*See Figure 5-40.*)

5-39. Using the nose of a chainsaw and a large chisel, cut mortises where the new log will join the old ones.

In the other ends of the logs, cut corner joints that match the style of those that were used to build the cabin. Lay a log across the sill logs, parallel to the cabin wall. Repeat, laying successive courses of logs and tying them to the cabin with large mortise-and-tenon joints as needed. Add a roof to the new room to match or complement the roof on the cabin, then fill the spaces between the logs with chinking.

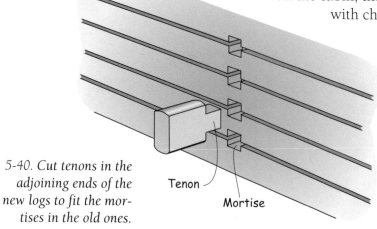

5-40. Cut tenons in the adjoining ends of the new logs to fit the mortises in the old ones.

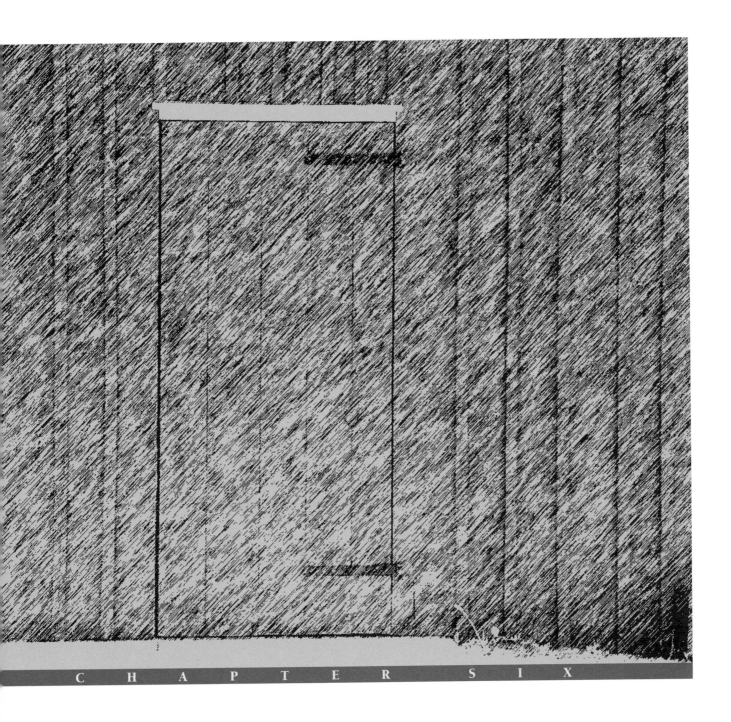

Siding and painting

*I*f you have an older building, you probably have siding problems.
The first surface on a barn or other outbuilding to show a little wear
is typically the siding. It's constantly exposed to the elements and has
nothing but a thin coat of paint to protect it. It shrinks and swells with
each change in the weather, causing the nails to loosen. From the
moment you spot the first bit of peeling paint or a loose board, it
becomes a constant chore to keep the siding in good repair.

Unfortunately, this is a chore that many of us fall behind on. Paint and siding problems seem minor compared to all the other repairs we have to do around a property, and so we put it off. The trouble is, while the siding problems are minor, they will lead to extensive structural damage if neglected for too long. The purpose of siding isn't just to enclose the building. It also covers the frame and protects it from excessive moisture. If the rain blows in through a hole in the siding and soaks frame timbers, they may start to rot. This is especially dangerous if the rain soaks an area where two or more timbers join together. The moisture will collect in the joints, the wood will rot, and the joints will be weakened.

Siding also keeps the wind out of a building. Wind can be more destructive than water. If the structure is missing siding boards on the *windward* side (the side that faces the prevailing winds in your location), the building will "fill with air" on a windy day, raising the air pressure inside. This trapped air presses out, putting stress on all the surfaces of the building, eventually weakening the entire structure. The air pressure may also blow off shingles and additional siding boards *from the inside*. (*See Figure 6-1.*) It is not uncommon for a strong wind to take a side or a roof off a building when there is a large hole in the siding. Years ago, I was part of a crew that was replacing the siding on a dairy barn. We stripped most of the old siding off the first day, then went home for the evening. That night, a front with high winds moved through the area. When we arrived at the site the next day, the wind had peeled off the barn roof and laid it back like the lid of a box.

Fortunately, fixing a siding problem is relatively simple — much simpler than repairing structural problems. Fixing siding is just a "search and replace" chore: Find the loose, rotted, or missing boards and replace them. Every few years, it becomes a "scrape and paint" chore to forestall the deterioration of the siding. How you replace a siding board will depend on the siding materials and the manner in which the siding is attached to the structure. How you paint or stain it will depend on how it was painted before and how long it's been since the last coat of paint was applied.

6-1. The roofing and structural problems evident in this barn probably began with a siding problem. Missing siding boards allowed the wind to enter the barn and lift the asphalt shingles from the inside. Once the shingles were lifted, the wind was able to loosen them and rip them off the roof. With the structure of the barn exposed to the elements from holes in the roof and siding, it began to decay.

Common types of siding

The siding on most older structures is made from wooden boards. Newer barns and out-buildings may be sided with sheet metal, vinyl plastic, plywood, and pressboard. These materials can be applied vertically or horizontally.

Vertical siding

Traditionally, barns and most outbuildings were covered with vertical siding for two reasons. (1) The post-and-beam structure of a barn was easier to cover vertically, and (2) it was more economical than applying horizontal siding. You can find several types of vertical siding on old barns.

Vertical boards — Boards attached vertically edge to edge with a small gap between them were typical used on tobacco barns and other structures in which ventilation was paramount. (*See Figure 6-2.*) The boards wear quickly with three sides of each exposed to sunlight, wind, and rain. In addition, such a structure is not weather-proof. The underlying structure is exposed to moisture and may be damaged. If you're converting a building with this type of siding to some other purpose, you'll want to replace the siding completely or upgrade it to board-and-batten siding.

Board-and-batten — This siding begins with vertical boards attached edge to edge. A layer of battens is then attached over the seams to weatherproof them. (*See Figure 6-3.*) As long as the battens are kept in good repair, the siding will protect the structure against the elements. If you have a building with structural problems in specific spots, you may find that the battens are missing in that area.

Vertical tongue-and-rabbet — As mill-work companies became more common, this wooden siding became more popular. Tongue-and-rabbet siding offered the same protection as board-and-batten siding, but it required fewer materials. It also added a great deal of strength to the structure. (*See Figure 6-4.*)

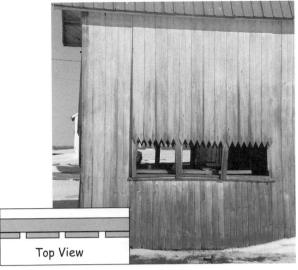

6-2. *Vertical board siding*

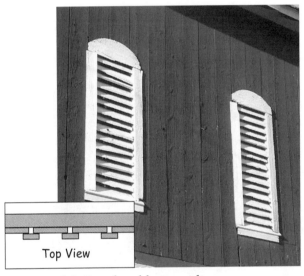

6-3. *Board-and-batten siding*

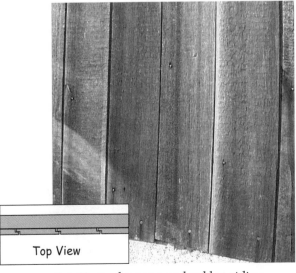

6-4. *Vertical tongue-and-rabbet siding*

Horizontal siding

When boards are attached horizontally to a structure, they must be overlapped to provide adequate protection against the weather. Consequently, the earliest forms of horizontal siding — weather boards and clapboards — required more materials than did vertical siding. Horizontal siding also required closely spaced frame members to properly attach them. The supporting members of post-and-beam structures are much too far apart to keep horizontal siding rigid. Consequently, you don't often find it on an old barn; it's more common on smaller framed outbuildings.

Weather boards — The earliest type of horizontal siding was made by overlapping ordinary boards. Often, these weather boards were not even trimmed at the edges. They were sawed "through and through" from the log so that the outside of the trunk became the edges of the boards. This made maximum use of each board and helped to make up for the lost materials where the boards overlapped. *(See Figure 6-5.)*

Clapboards — Clapboards were a refinement of weather boards. Where the boards overlapped, the material could be quite thick. Clapboards were sawed with a wedge-shaped cross section and were installed with the narrow edge of the wedge facing up. *(See Figure 6-6.)*

Horizontal tongue-and-rabbet — Like its vertical counterpart, this type of siding came into use with the growth of the millwork industry. It's installed with the tongue up. The rabbet laps over the tongue and makes a weatherproof seam. *(See Figure 6-7.)*

6-5. Weather board siding

6-6. Clapboard siding

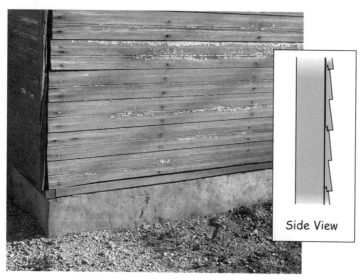

6-7. Horizontal tongue-and-rabbet siding

Pressboard siding — Developments in "manufactured" wood products produced this light, relatively inexpensive horizontal siding made from compressed wood fibers. In many respects, pressboard siding is a modern version of weatherboards. The type of pressboard mostly commonly used to side outbuildings is a simple board with little or no taper from edge to edge. *Pressboard siding must be painted* to be properly protected from the elements.

Its primary advantage over solid wood, aside from the lower cost, is that it does not "weather" like wood, which develops ridges and valleys where there are bands of springwood and summerwood. Consequently, pressboard stays and remains attractive. Its smoothness also helps pressboard to shed rainwater. The material's disadvantage is that it has little structural strength. If you plan to replace the old siding on an outbuilding with pressboard, you may have to reinforce the frame considerably so as not to weaken the overall structure. You may be better off simply applying the pressboard over the existing wood siding. (*See Figure 6-8.*)

Aluminum and vinyl siding — Aluminum and vinyl sidings are purely decorative coverings. They are molded in strips that mimic the appearance of clapboard. They are less expensive than wood siding and somewhat easier to apply. They don't require paint — the colors are baked onto the aluminum or mixed into the vinyl. However, you can use special paint if you ever want to change the color to give a building a fresh look. Aluminum and vinyl siding have no structural strength whatsoever, so the same consideration applies to these materials. Either reinforce the frame or apply them over the existing siding. (*See Figures 6-9 and 6-10.*)

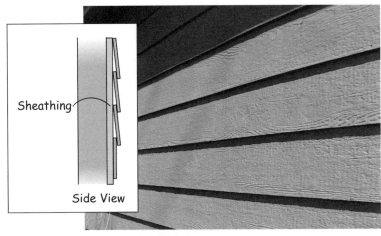

6-8. Pressboard siding

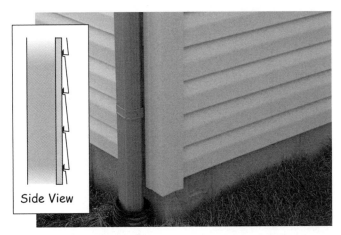

6-9. Aluminum siding

6-10. *Both vinyl and aluminum siding are siding "systems" — they come with special hardware for joining strips and attaching them to the structure. There is also a variety of fittings to trim around windows, doors, vents, and corners. If you use this type of siding, you should also use the system hardware and its trim.*

Top View

6-11. T-111 Plywood siding

Sheet siding

Almost all the new barns that I have seen going up in the last few years are sided with sheet materials. Sheet siding not only saves money over solid wood, it also saves time and labor. The type of sheet goods used depends largely on the size of the building.

Plywood sheathing — Small storage sheds and other outbuildings up to the size of a garage often are sided with exterior grade plywood. This plywood is usually milled so that one face resembles vertical board siding. The edges are cut with interlocking tongues and rabbets so that they overlap, making watertight joints. Plywood is an effective and relatively inexpensive replacement for damaged exterior walls. By the way, if you want to use the proper term at the contractor's desk, plywood barn siding is referred to as T-111. *(See Figure 6-11.)*

Metal sheathing — Metal sheathing has rapidly become the preferred siding for outbuildings, and with good reason. Besides being relatively inexpensive and easy to install, it's an excellent barrier from the elements and requires little maintenance. Modern metal sheathing is made from aluminum that is impervious to the elements and has the color baked on, so it never needs painting. Older metal-clad barns may have been sided with tin or galvanized steel, which rusts if not kept painted. *(See Figures 6-12 and 6-13.)*

Top View

6-12. Modern aluminum siding is often an excellent choice for restoring older, wood-sided barns. The vertical ridges in the sheet aluminum preserve the look of traditional barn siding.

6-13. Older galvanized steel siding

Finding and fixing siding problems

As a young man, I once served under several experienced barn builders on a construction crew resurrecting an old Shaker barn. The first step to any barn restoration is identifying the problems, so when it came time to inspect the siding, the crew trooped inside.

"What are we looking for?" I asked one taciturn old foreman.

"Daylight," he replied.

When properly sided, the wall of a barn or an outbuilding should be relatively weatherproof. Unless you're working in a tobacco barn or a corn crib, which require ventilation, if you can see daylight through the siding, you'll soon have a siding problem. You might have one already.

Inspecting the siding

Actually, inspecting the siding is a little more complicated than looking for daylight. But just a little. Here's what to look for:

Rotting ends — Siding boards with rotten ends is one of the most common problems you will encounter on any board-sided barn. (*See Figure 6-14.*) The ends of a board act like sponges when moisture is present. Unless painted, the ends of boards will soak up moisture. The moisture creates a fertile environment for decay-causing bacteria, and the wood begins to rot. Look hard at the boards around windows. Where the rotted end of a board meets a sill, chances are the sill is damaged, too. Plan on replacing any rotted wood you find.

Cracks and splits — Cracks and splits in wood siding collect moisture. The cracks will grow and the wood will begin to rot if these are not repaired. (*See Figure 6-15.*) The simple solution is to caulk any crack and paint it. Where the wood has actually cupped away from the crack, you may have to cut out a section of the board and replace the wood. If a crack near a structural member is left unrepaired, it may lead to interior rot of the member. When cracks and splits are found in rabbet-and-groove siding, you may have to cut out a section of that siding to replace a damaged piece.

6-14. The bottom ends of siding boards tend to rot long before the rest of the board because moisture runs down the board, then wicks up into the ends. You can prevent this by keeping the ends painted.

6-15. Siding boards often crack or split along the grain. Again, the culprit is moisture — water soaks into the wood, freezes, expands, and splits the board. Painting prevents this.

6-16. Replace missing battens as soon as possible to prevent water from reaching the structure of the barn.

Missing boards or battens — When a batten falls off, rain and moisture penetrate between the boards it was protecting. This leads to deterioration. To find all the damage, don't just look at the outside where the board or batten is missing. Inspect the inside, too — especially the frame members where the missing piece was attached. *(See Figure 6-16.)*

Protruding nails or enlarged nail holes — Wood expands and contracts with changes in moisture content. As a board moves, it can actually work its fasteners loose, pushing the nails out so they protrude. Once this happens, the exposed metal comes in contact with the weather. This causes the nails to rust and the surrounding wood to decay.

Peeling paint — If the paint is peeling, the rainwater is reaching the wood. This, in turn, will start the wood to decay. The *springwood* will decay more quickly than the harder, denser *summerwood,* and the barn siding will develop a rough "weathered" texture. Once this happens, the siding requires more paint to cover it, and the paint is much more difficult to apply.

EXTRA HELP

What gives barn siding its weathered look? Each year, a tree goes through a cycle of growth that produces wood cells of different density. during the spring, the tree grows very quickly. This *springwood* is softer and less dense than the *summerwood* that's created later in the growing cycle. The layers of springwood and summerwood appear as annual rings when you crosscut the wood, or as light and dark bands when you rip it into boards. As the siding ages, the springwood wears and weathers away much faster than does the more dense summerwood. The siding eventually takes on a rough texture that exaggerates the grain of each board.

New wood

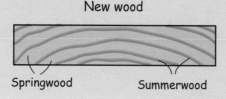

Springwood   Summerwood

Weathered wood

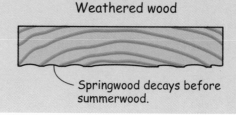

Springwood decays before summerwood.

Caulking cracked and split boards

Small cracks can be plugged easily with caulk. Use a caulk that is compatible with the paint you will later use to cover the repaired siding. some silicone caulks, while they are more durable than other materials, will not take paint at all. Common painter's caulk works best for most repairs. If you are not sure whether your caulk will be compatible with the paint you plan to use, ask at your local paint or building supply store. When caulking, force the caulk into the cracks or splits, then remove any excess material by smoothing it even with the surface of the board with a putty knife or paint scraper. (*See Figure 6-17.*)

6-17. When caulking a cracking or split board, lay down a bead of caulk then smooth or "wipe" the caulk with a putty knife to force it into the crack.

Replacing rotted wood with fiberglass

Small rotted parts can be replaced or repaired with a fiberglass patch. This is a good fix when you are working on a molding or profiled piece of wood that has minor damage and would be hard to replace. Clean the area you want to repair, apply the fiberglass, let it harden according to the manufacturer's instructions, then sand it level with the surrounding surface. Because the fiberglass patching material is expensive, this can be a costly repair. It should only be used for small areas, or when you can't restore the rotted sections any other way.

Replacing missing battens and trim

Using an original batten as a template, make new battens to match the width and thickness of the original. When replacing only a section of a batten, splice the new wood to the old with a *scarf joint.* Cut the adjoining ends at a steep angle so the water will run out and away from the seam, should it penetrate it. Attach the battens with galvanized nails. Caulk any scarf joints and smooth the caulk.

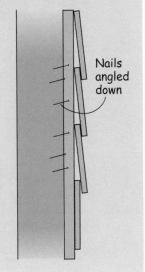

EXTRA HELP

To make your nails hold better, drive them at an angle, rather than pounding them straight into the wood. Make gravity work for you — angle each nail down slightly so that the weight of the board will tend to pull it tight against the wood to which it is fastened. Also angle the nails left and right and vary the angle with each nail. This "hooks" the board into the wood, making it harder to pull out.

Nails angled down

Replacing rotten or missing siding boards

If the siding has deteriorated and some boards are missing, or specific boards need to be replaced, you need to plan each repair. How you replace a siding board has a great deal to do with the type of siding. For simple sidings like vertical boards and board-and-batten, it's just a matter of removing the nails, discarding the old boards, and nailing up new boards in their place. For other types, the siding is like a jigsaw puzzle that you must disassemble in the proper order.

Vertical siding — To replace vertical boards, first remove the battens and any trim covering them. *(See Figures 6-18 through 6-20.)* Be careful when removing the trim, as you will probably want to reinstall it when you have made your repairs. Typically, this old trim will be wider than dimensional trim boards you can buy today, and it can be expensive to match it with newly milled wood. After removing the battens and the trim, use a cat's paw to remove any nails that remain in the boards. Although it was once considered thrifty to save, straighten, and re-use old nails, this is no longer the case. In the long run, it's wiser to replace them with new, especially if the old nails are starting to rust. If you are trying to save part of a board, or if nail heads are popping off and frustrating your attempts to remove a board in one piece,

*6-18. The best tool I've found for removing old siding boards is a **cat's paw**. To use this tool, place the paw beside the nail with the claws angled down slightly, Strike the paw with a hammer to drive the claws under the nail head.*

6-19. Pull back on the handle of the cat's paw, using it like you would a claw hammer or a crowbar to remove the nail.

6-20. If you can't get nails loose with a cat's paw, try a reciprocating saw. Pry the board loose, then sneak in behind it with a flexible metal-cutting blade and cut off the nails.

consider using a reciprocal saw. You can pry the board loose and cut off the nails using this saw and a flexible blade.

Clapboard and tongue-and-rabbet siding — Replacing a clapboard or a tongue-and-rabbet siding board takes just a little more forethought. If you replace just part of a board, you must match the shape and slope. If you replace one or more boards, you must remove nails and siding in the proper order because the siding pieces actually lock in place or overlap each other.

Spot repairs. To replace just part of a siding board that is damaged or rotted, remove any trim that's in your way. Drill one or more holes through the affected piece of siding to allow you to insert the blade of a jigsaw or reciprocating saw through the siding. Carefully cut out the damaged section. If you can, it's a good idea to remove material frame member to frame member. This will make it easier to fasten the repair in place.(*See Figures 6-21 and 6-22.*)

Using the section you just removed as a template, cut a repair piece to fit the gap. If needed, nail cleats to the frame members to provide an anchor for the repair. Nail the repair piece in place and caulk the seams between the old wood and the new.

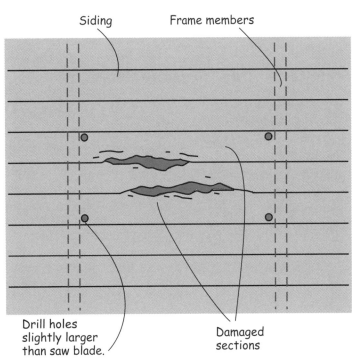

Siding   Frame members

Drill holes slightly larger than saw blade.

Damaged sections

6-21. To remove just a section of a siding board, you must make a piercing cut with a saber saw or a reciprocating saw. First drill a hole where you want to start the cut.

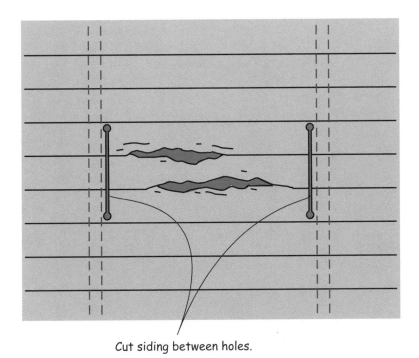

Cut siding between holes.

6-22. Insert the blade in the hole and cut out the section. Keep the cutout and use it as a template to trace the shape of the repair piece.

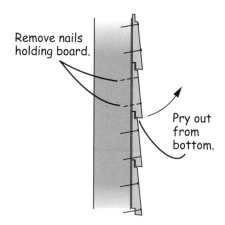

Remove nails holding board.

Pry out from bottom.

6-23. *To remove a single piece of clapboard, you must remove the nails in the board and the lower nails in the board above it. Where the boards overlap, the nails are typically driven through both boards.*

Replacing entire boards. To replace an entire board in the middle of the exterior wall, remove the trim. Then remove the nails that hold the board in place. *(See Figure 6-23.)* Loosen (but don't remove) the board above it by sliding a small, flat pry bar under the bottom edge and prying it up an inch or so. Slide the old board out and use it as a template to cut a new one.

Slide a new board in place. Tap the board above it (the one you loosened) back down. Nail through the bottom of the overlapping board above and through the top of the replacement board. Also nail the bottom of the replacement board. Prime and paint the new wood, and your repair will be almost unnoticeable.

If you're working with old clapboard, positioning it requires some knowledge of long-ago methods and a special tool that you can pick up at a flea market — a clapboard gauge. When clapboard was applied to the side of a building, it was commonly spaced wider at the top to combat the optical illusion of boards getting narrower as they go up. This also made buildings look taller, satisfying the vanity of farmers of the past. *(See Figure 6-24.)* Carefully measure the spacing when you remove the old clapboard and be sure to duplicate it when you repair or replace it.

6-24. *When clapboards are properly installed on an old building, the exposed width of the clapboards should increase the higher the boards are on the building. This makes them appear to be the same width from eye level to the top of the building.*

Using a clapboard gauge

To properly space horizontal lapped siding and clapboard siding on an old building, you need a *clapboard gauge*. This is a marking gauge with preset "steps." You use it to mark the *exposed* width of a piece of clapboard (the part of the board that shows when the siding is installed).

1 Most clapboard gauges were homemade. You can find them at flea markets or simply make your own. Cut the steps in the base in 1/4-inch increments. For a scribe, drive a #8 screw through the bar, near the top. The distance between the scribe and the last step on the base should be 1 inch less than the width of your clapboards.

2 To use the gauge, first determine how much of the clapboard you want to show. This will determine which step you hook over the bottom edge of the clapboard. Pressing the gauge against the wood firmly, draw it along the length of the clapboard, scratching a line.

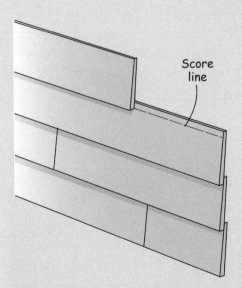

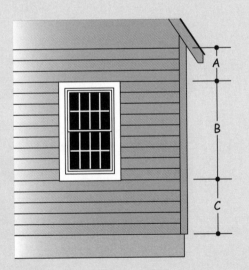

3 Nail the next piece of clapboard in place, aligning the bottom of the board with the scribed line. Then repeat the process, scribing another line. Remember that the revealed portion of the clapboards should grow wider as they go up. However, you should always overlap the boards by at least 1 inch.

4 Old-time carpenters also aligned the horizontal lines of the clapboards with the tops and bottoms of windows and doors. This makes the doors and windows seem more integrated with the walls and present a more pleasing appearance. To do this, however, takes some careful planning. Measure the distances A, B, and C before you start so you can figure how to space the clapboards over each section.

6-25. In many cases, it's advisable to apply new siding over old. The old siding not only serves as a substrate for the new, it's an important part of the structure. If you remove it, you may weaken the outbuilding. The aluminum siding on this barn-turned-office building was applied directly over the old vertical boards.

Installing new siding

If the siding is beyond repair, or if you want to save maintenance costs over the long haul, install new siding. There are new products on the market today that preserve the traditional look of old barn siding, but are much easier and less expensive to keep up.

Determine whether you want to strip off the old siding or apply the new siding over the old. If you decide to remove and replace the siding, you may have to brace the structure internally before stripping it off, especially if the framing members are spaced more than 24 feet apart vertically. On these structures, the siding acts as a structural member. (*See Figure 6-25.*)

Installing sheathing

Depending on the type of siding you choose, you may need a substrate for the new siding. Piece in boards or plywood if you are using your old siding as a substrate for the new. If not, install sheathing over the frame members.

If you are working on a post-and-beam structure, you may need to add nailers to attach the new sheathing or siding. You don't want your framing members (or nailers) to be more than 24 inches apart. Check out the manufacturer's instructions and requirements if you are using plywood sheathing, metal sheathing, or aluminum or vinyl siding. (*See Figure 6-26.*)

6-26. When siding most frame structures, the siding is attached to the vertical studs. On post-and-beam and pole structures, the siding is nailed to horizontal components such as these nailers.

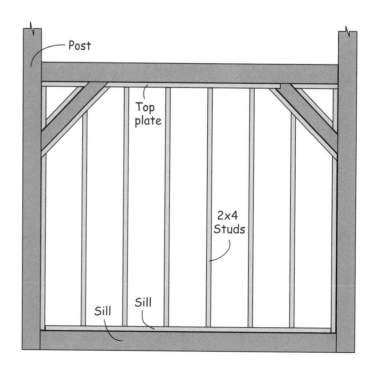

Installing a vapor barrier

Once you have installed the sheathing, add a vapor barrier before applying new siding. A vapor barrier allows moisture to escape from the structure without letting moisture in. Kraft paper and tar paper are the traditional favorites for a vapor barrier. More and more, builders are using housewrap. It's tear resistant, unlike tar or kraft paper; it is made almost wholly from recycled materials; and it does not break down as it ages. It's easy to install and comes in rolls 10 feet long, allowing for quick installation.

Installing vertical boards

When siding with vertical boards, make sure you place them at least 8 inches above the ground. The ends of boards act like sponges, so I recommend that you paint the ends prior to installation. Hang the first board perfectly plumb (choose your straightest board for this) and check that subsequent boards are plumb. If you notice that your boards are getting out of plumb, make slight adjustments as you work.

Boards should be placed ¼–½ inch apart to allow for expansion. (See Figure 6-28.) Drive galvanized nails at an angle to install the boards, and make sure the annual rings on the ends of the boards cup out. (See Figure 6-29.)

After the boards are up, apply battens to the seams using nails driven at an angle. Install any trim on the ends. Caulk any gaps in the seams before painting. You don't need to caulk if your seams are tight or if you are priming and painting the wood soon after installation.

6-27. Housewrap is a special paper that serves as a vapor barrier. It's a good idea to apply this over the old siding before you install the new.

EXTRA HELP

You can snap level lines on your framing to act as reference points to help you keep the boards level as you install them. This will prevent you from having to check each board for levelness; you can simply measure from the top and bottom of a board to the nearest reference line.

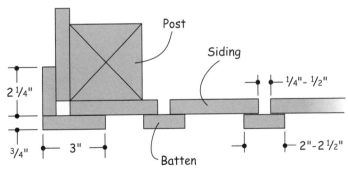

6-28. Space the siding boards ¼–½ inch apart to allow for expansion and contraction. Trim the corners with wood strips to protect the post and framing members from the weather.

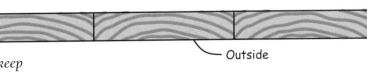

6-29. Which side of a siding board should face out? The heartwood or inside of the tree should face out so the annual rings cup out as you look at the ends of the board. Boards tend to cup in the opposite direction of the rings, so this will keep the edges of the boards flat against the frame.

Installing lapped board and clapboards

Consider prepainting your clapboards. Prepainting will help to stabilize the clapboards. It is always better to paint both sides of any board used in an exterior application to keep it from cupping. (*See Figure 6-30.*) Install the boards in a lapped fashion, starting at the bottom. Begin with a nice, straight board and install it perfectly level, making sure to keep the bottom edge of the board at least 8 inches above the ground.

Nail clapboards at the top of the board on the wall nailers, studs, or framing members no more than 24 inches apart on center. Keep the spacing of the siding boards consistent until you reach eye level. At that point, use the clapboard gauge to position the boards so the spacing increases slightly with each successive course. (See "Using a clapboard gauge" on page 145.) When you reach the top, trim the last clapboard in place and cover it with 1x trim, completing the job. Install side trim and enjoy the view.

6-30. By painting both sides of siding boards before installation, you will greatly reduce the curling and cupping that will occur when the boards are out in the weather.

EXTRA HELP

In addition to using real nails, you can also use a construction adhesive. This reduces your workload because it attaches the batten and caulks the seam at the same time.

EXTRA HELP

Make a clapboard hanging jig to simplify the task of hanging your clapboards evenly. You can adjust the exposed width of the clapboard by using shims with the jig. Simply insert an equal number of shims as you go up to decrease the exposed width of successive clapboards.

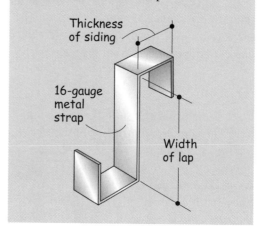

Thickness of siding

16-gauge metal strap

Width of lap

Painting

Would you like to know why barns are usually painted red? Because, when it became all the rage to paint your barns in the late 1800s, that was the least expensive color available. Before that barns and outbuildings were left to the weather — painting was an extravagance.

It's still considered something of an extravagance, if all the barns and outbuildings in sore need of paint are any indication. But over the years, paint has been developed to last longer and provide better protection than it ever has been before. A good coat of paint not only improves the looks of an outbuilding, in the long run it will save you time and money on maintenance.

Repainting old wood siding

Inspect the siding for any warning signs of moisture soaking into the wood before you begin preparing the surface for paint. *(See Figure 6-31.)* If there are areas where the paint is peeling off but others where it is adhering well, look for evidence of a moisture problem in the peeling areas. Are the bottoms of these affected boards in contact with the ground? Is there a leak in a downspout just above the affected area? Probe the wood with a screwdriver. Is the wood punky?

Also inspect the siding to see if there is just too much paint on the boards. Thick coats of paint on the siding will eventually flake off. This is because the paint restricts the boards' ability to breathe with the changes in relative humidity. All paint is semipermeable — it allows moisture to pass through, albeit slowly. As the humidity falls and the wood loses moisture, the moisture must be able to escape through the paint. If the paint is so thick that it has become impermeable, the moisture literally pushes the paint off of the building.

If paint isn't too thick and retains its permeability, then you must choose a paint that will bond to the old coat. Take a sample board to your paint supplier to determine how many coats of paint are on the building and what type of paint will adhere properly to the old. *(See Figure 6-32.)*

Removing old paint — If you determine that the paint on the building needs to be removed, there are different tools and techniques you can choose from.

6-31. *Paint peels because it loses its elasticity over time and no longer expands and contracts with the wood. It begins to delaminate, splitting off from the wood surface. Once this begins to happen you must scrap off the old paint and apply new.*

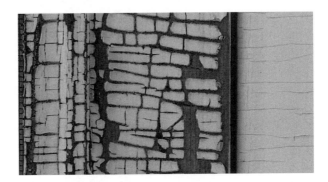

6-32. *Occasionally, the paint blisters instead of peels. It delaminates from the wood but doesn't split and peel off. This happens when there is a moisture problem, perhaps caused by a leaky roof or missing battens. The moisture inside the building builds up and "pushes" the paint off the wood from the inside out.*

Pressure washing — Pressure washing machines are great for removing old paint and dirt from siding. However, you must be careful not to damage the surface of the wood. Just like normal wear and tear removes springwood faster than summerwood from a board, a pressure washer can actually gouge out the springwood from a board face. I almost ruined a paint job by improperly using a pressure washer for prep work.

Corn cob blasting — "Sandblasting" the siding with ground corn cobs instead of sand is an effective way to remove old paint. The paint has to be fairly loose to begin with, i.e., this method is more effective when the siding has not been painted at all for many years. The ground cobs act as a soft abrasive, breaking up the paint and forcing it off of the siding. The only drawback is the expense of renting the equipment. (*See Figure 6-33.*)

Power scraping — There are power scrapers on the market that are specifically made for lap siding paint removal. These work on the same principal as a grinder except that the revolving disc is actually a carbide-tipped blade that scrapes the paint off of the surface. You are left with a bare wood surface that you can then prime and paint. (*See Figure 6-34.*)

Hand scraping — No matter what power method you use to remove most of the old paint, you are going to have to do some hand scraping to get at stubborn and hard-to-reach areas. I recommend you purchase a "two-fisted" scraper — one with a long arm and handles for both hands this gives you more leverage. (*See Figure 6-35.*)

Final preparation and painting — Once you have all of the loose paint off, go back over the siding and caulk any gaps or butt joints that will fill with water when it rains. Doublecheck trim boards, windows, and doors. Once you have the siding sealed up, prime it with a quality primer. If you are painting water base as your final coat, use a primer that will bind old paint with the new water based paint. Your paint supplier can help you determine what you will need.

When the wood is primed, paint the siding. Because the wood is older, dryer, and rough, this will probably take longer than painting new wood. Apply the paint in a thick coat, thoroughly "wetting" the wood. The paint will give the best protection if the coat is continuous with no voids or gaps. If you come to a crack or a gap that you haven't filled, stop and caulk it.

TO BE SAFE

Never use heat guns or propane heat to strip off paint from barn siding. This is a terrible fire hazard. You may start a fire on the other side of the board and not even know it. Wood can smolder for several hours with no outward signs. Also, heating up aged paint creates vapors that you don't want to breathe. It is just a bad idea all around to try to remove barn paint with heat.

EXTRA HELP

Apply a double coat of paint to any exposed end grain. The end grain at the end of a board soaks up moisture much faster than the "flat grain" on the edge or face. An extra coat helps to seal the end grain.

6-33. Perhaps the kindest, gentlest, and fastest way to take old paint off a building is to blast it with ground-up corn cobs.

6-34. A power scraper removes the paint with a rotary blade. This requires some skill; it's easy to let the blade eat into the wood.

6-35. If you have a considerable amount of hand scraping to do, use a two-handed scraper. It's faster and less tiring than the traditional one-handed variety.

Choosing exterior paints and coatings

Whether you are painting an old surface or covering a new one, there are many different paints and coatings for you to choose from. A painting specialist in a paint store or building supply center will help you choose the best paint for your particular job, but here's a brief summary of the available materials.

Paint

Paints are opaque — you can't see the wood grain beneath them.

Latex flat — Fast drying, odor free, thinned and cleaned with water. More permeable than other finishes — lets the wood "breathe" moisture in and out with changes in relative humidity. Will not adhere to old "chalky" painted surfaces or those coated with alkyd paint.

Alkyd flat — Fast drying, slight odor, thinned and cleaned with mineral spirits. Adheres well to chalky paint and all other surfaces except masonry and metal. Blisters more easily than latex in humid conditions.

Oil flat — Slow drying, distinct odor, thinned with linseed oil, cleaned with mineral spirits. Once the mainstay of exterior coatings, oil paints have all but been replaced by more durable latex and alkyds. However, you may want to use them for historic restorations.

Glossy trim — Available in both latex and alkyd. A good choice for locations which collect dirt and need to be cleaned occasionally, such as door and window trim.

Marine paints — These use extreme durable epoxy, urethane, or alkyd resins and are longer wearing than all other types of paint. But they are also much more expensive. However, marine paint may be worth the expense if you live in a location with harsh weather.

Porch and floor paints — Formulated to withstand heavy wear-and-tear from abrasion-resistant alkyd, latex, epoxy, or urethane resins. There are also rubber-based floor paint for extremely wet locations. These adhere to wood readily, but you must specially prepare concrete by scraping it with a sharp instrument to give it some "tooth."

Shingle paint — Extremely permeable to let siding shingles breathe. Works well for all wood shingles. Asbestos-cement shingles require a special paint, and creosote-treated shingles cannot be painted at all.

Metal paint — Formulated to prevent rust and corrosion, available in oil- and alkyd-based. These are used to paint metal siding and roofing. They usually don't require primers, but the surfaces must be clean and corrosion-free.

Stains and Clear Finishes

These are semi-transparent or clear, letting you see the wood grain beneath them.

Exterior stain — Formulated with the same resins as paint, but these either have less pigment or they substitute dye for pigment to allow the hardened coating to transmit light. These have the same properties and durability as paints with the same resin base.

Varnish — Three types available: alkyd, urethane, and spar. Alkyd lasts longer under normal circumstances, urethane is tougher when it must stand up to heavy use, and spar provides better protection against salt water if you live near the shore.

Wood preservative —Protects against damage from fungus and insects, and not much more. Some varieties are formulated to waterproof the wood.

Water repellent — Usually has a silicone base, makes the wood almost completely waterproof and preserves its natural appearance. Cannot be painted over.

Painting new wood

As when painting old wood, you must prime the new wood. If you are painting pressure-treated wood (so-called "outdoor" lumber), it's best to let the wood stand through one or two rain storms before applying primer or paint. This washes off a chemical residue on the surface of the board. It also gives the board a chance to dry and reach the same level of moisture content as other boards in the structure. Pressure-treated wood typically has a much greater moisture content when you first purchase it — this is why it's so heavy.

When applying paint, brush with the grain — the paint will flow on more easily with fewer brush marks. You will always have to do some brushing against the grain, but always try to finish up by brushing with it. When painting boards that lap one another, such as clapboards or board-and-batten siding, paint the edges first, then cover the faces. (*See Figures 6-36 and 6-37.*) If you paint the edges last, you're more likely to get runs and drips.

Painting aluminum and vinyl

Ask any professional painter and they will tell you that anything can be painted with excellent results. You simply have to know how to prepare the material correctly and what type of product to use to paint it. I recently painted vinyl siding with a result that is nearly impossible to distinguish from the color manufactured into the plastic originally. There are paints specifically manufactured to paint all types of exterior surfaces. Check with your paint supplier to find out what is available and what application best suits your particular situation.

6-36. *When painting clapboard with a brush, paint the bottom edges first. Dip just the tip in the paint and coat about 3 feet of the edge. Cover several edges.*

6-37. *When the edges are covered, switch to the faces. Brush with the grain in short strokes, then finish off in longer strokes to smooth the paint and eliminate brush marks.*

Painting tools

Painting doesn't have to be done with a paint brush. There are several better alternatives when you have a large surface (such as the sides of a barn) to cover.

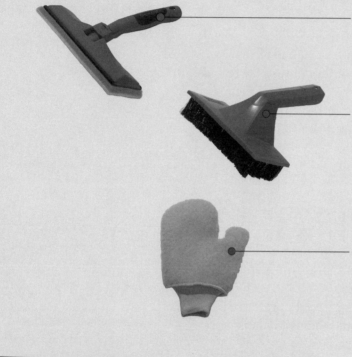

A *pad applicator* is particularly good for uneven surfaces as well as wide, flat ones. The foam rubber backing on the pad conforms to any contour.

A *rough surface paint brush* has stiff, stubby bristles to help work the paint into the crevices of a rough-texture surface. It's handy for rough-sawed wood, weathered wood, concrete block, stucco, and brick.

A *mitten applicator* is a speedy way to paint portions of buildings with lots of small surfaces such as louvers, vents, railings, decorative wood, and metal work.

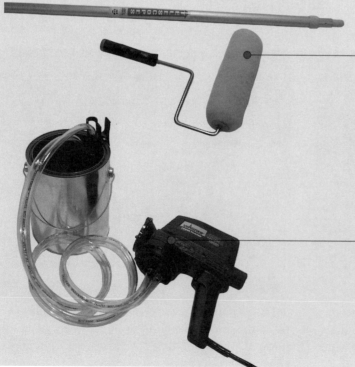

A *long-nap roller* and a *telescoping extension* reach the high spots. The long nap get the paint into cracks and crevices, and the extension cuts down on the time you have to spend moving the ladder.

An *airless spray gun* covers a variety of surfaces quickly and easily. But to prevent clogging, you often have to thin the paints. This reduces the thickness of each coat and often makes it necessary to apply additional coats for adequate coverage.

Interior walls

There are two types of interior walls: load-bearing walls, which are discussed in Chapter 5, and non–load-bearing walls, referred to as partitions. The structural requirements for partitions are really quite minimal. The partition framing must hold up the wall covering, typically gypsum board, and support it at close enough intervals that it doesn't crack in normal use, or abuse. These requirements are adequately met by studs spaced every 16 inches.

The framing of a wall must do more than meet structural requirements, however — it must also provide backing for all of the trim work and the multitude of things that we fasten to walls, things like cabinets, light fixtures, towel bars, clothes bars, and on and on. In kitchen or bathroom walls these "backers" or "nailers" can be more time-consuming to fit and install than the structural elements. But omit them at your peril — few things about finish work are more frustrating than discovering that there's only one suitable place for a particular fixture and that all you have to fasten it to in that place is a half inch of gypsum board.

Once the partition framing is up and all the necessary backers are installed in both the partitions and the load-bearing walls, you can get on with the wiring and plumbing (covered in Chapter 10), then the insulation, and finally the wall covering.

Adding partitions

Renovating or adding to an old structure differs from building a new structure in two important ways. First, very little about an old structure is level or square, or even straight for that matter. And second, you can't go about the work in the efficient sequence typical of new construction —

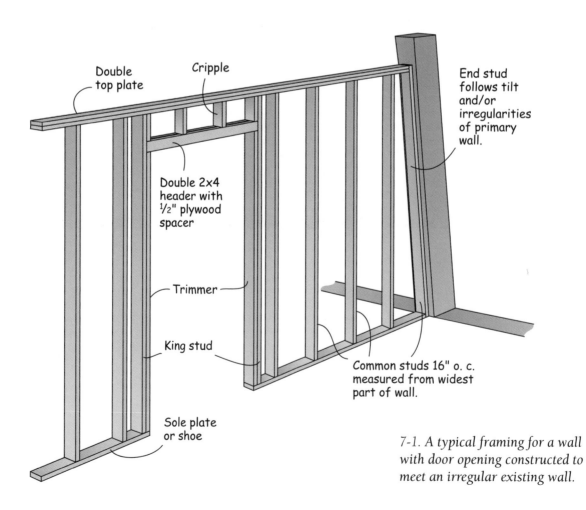

7-1. *A typical framing for a wall with door opening constructed to meet an irregular existing wall.*

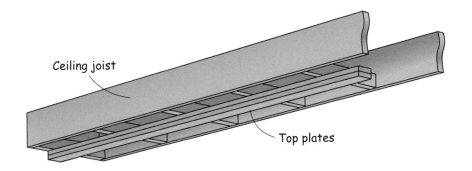

Ceiling joist

Top plates

7-2. *To attach a partition to a ceiling between ceiling joists, fit 2x4 nailers on 16-inch centers between the joists. For a partition directly under a ceiling joist, you will have to fit 2x4 nailers between the joists in both adjoining bays.*

instead you often find yourself adding to structure that wouldn't be there yet in new construction. When these two differences both conspire against you, you can find yourself erecting a partition in a space where neither top nor bottom is either level or straight and the two sides are equally impossible to measure. Don't succumb to the temptation to go out and mow the lawn; you have some simple and very effective ways to deal with the problem.

First, before you begin, put aside any intention you may have had of using the technique common in new construction: assembling the framework of a wall flat on the floor and then lifting it up into position. That technique works just fine on a newly built floor platform with no ceiling joists yet in place. But even in the unlikely event that both your floor and ceiling are level and flat, you still won't be able to raise the framework up into position unless you make it a substantial fraction of an inch shorter than it needs to be. And if you do make it shorter, you'll then have to add shims between the top plates and the ceiling joists. Adding those shims will use up any time you may have saved by assembling it on the floor. You're better off building it in place.

Start by preparing the existing ceiling structure. If the partition is to be perpendicular to the ceiling joists, this may require no more than marking the location of the partition with a chalk line. If the partition is to be parallel to the joists, you'll first have to fit nailers every 16 inches between ceiling joists in the bay where the partition is to be. Make sure you keep the nailers flush with the bottom of the joists. If the location you've chosen for the partition falls directly under a joist you'll have to fit nailers in both of the adjoining bays. Keep in mind that the nailers are not just something to nail the top plates to; they also provide backing for the ceiling material, which is typically gypsum board. With the needed nailers in place, mark the location of the partition with a chalk line.

Now mark the location of the partition on the floor, directly under the marks on the ceiling, with the aid of a plumb bob and a chalk line. You'll find a plumb bob much easier to use in a situation like this than a level. (The plumb bob is also virtually indestructible; the level is definitely not.) If the partition is perpendicular to the joists, the floor needs no further preparation. If it is to be parallel to the floor joists and the flooring is less than heavy planking, you should fit nailers between the joists as you did at the ceiling, but up tight against the flooring instead of flush with the bottom of the joists.

Joining partitions to existing walls

Next, decide how your partition will join the walls you already have. This will depend largely on the structure of the existing walls. If they are framed with dimension lumber as your partition will be, you've got three choices: You can add two additional studs with blocking between, which is how they did it when wood was cheap and heating oil was nine cents a gallon; you can add horizontal nailers just as you do between ceiling joists parallel to the wall; or you can add plywood flanges to the back of the end stud to carry the wall surface on the wall you're joining to.

If your partition will join a post of a post-and-beam frame you need only to lay out its location with a chalk line. You'll simply nail the end stud to the post, though if it's 200-year-old hard maple nailing it won't be particularly simple unless you use hardened-steel, spiral flooring nails.

The best way to join to a stone wall that does not have a framed wall inside it for insulation and utilities is to put up the end stud first, then point the voids between the stud and the wall just as you would point between stones in the wall. In this case, wait until the plates are in place and then staple roofing felt to the surface of the end stud that the pointing mortar will contact. It's also wise to drive nails into the stud so their heads project into many of the larger voids. This gives a bit of mechanical connection between the partition and the stone wall and could help prevent the end stud from bowing one way or another.

7-3. Early framing methods called for two additional studs with blocking between when joining one wall to another. The two studs carried lath and plaster, and blocking provided a fastening surface for the joining wall. The whole structure was virtually impossible to insulate, and wiring had to pass through holes drilled in the two added studs.

Blocking

Nailers

7-4. Today's gypsum board walls are quite adequately supported by nailers between common studs. This allows room for insulation outside the nailers. Wiring can pass through without additional holes.

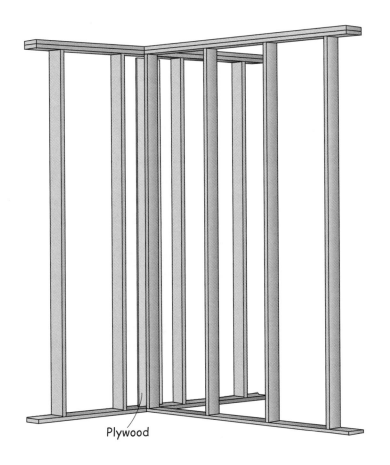

Plywood

7-5. An even more efficient arrangement uses 6- to 8-inch-wide strips of ½-inch sheathing plywood nailed to the back surface of the end stud of the joining wall before nailing the end stud in place. Gypsum board is easily screwed to the projecting plywood. An additional inch of insulation room is gained, and often the plywood used is leftover scraps.

Joining a framed partition to a log wall is probably the most time-consuming situation you could be facing. Assuming that it is a genuine old log outbuilding and not a recently manufactured log building, most of which are built for residential use, the building probably has considerable historical value and you should not cut into the logs to attach your partition. Instead, frame the rest of the partition first, then temporarily nail up two pieces of 2x of appropriate width as shown in *Figure 7-6*. Scribe the shape of the wall surface onto the 2xs, saw to the scribe marks, and toenail them to the plates with a ½-inch shim between the two shaped pieces. You'll also have to scribe and fit your partition surface material when you get to that stage.

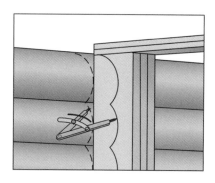

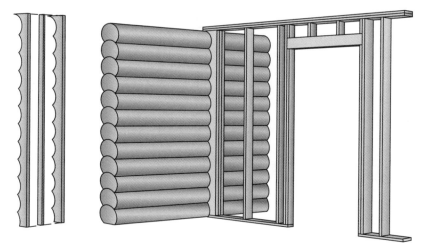

7-6. To fit a partition to a highly irregular wall like one made of round logs, first put up the plates. Then temporarily nail two pieces of 2x of appropriate width between the plates (left). Position them plumb and up against the irregular wall. Then, keeping the points of a large pair of dividers level, scribe the shape of the wall onto the two pieces of 2x. Remove the temporary nails and saw to the scribe marks with a portable jigsaw. Sandwich a ½-inch shim such as sheathing plywood between the two 2x pieces to make them up to the thickness of the partition framing (middle). Fit them back between the plates, snug to the irregular wall, and toenail them in place (right). When installing the wall surface, be careful not to split off the projections of the scribed pieces. Use drywall screws rather than nails if the wall surface will be gypsum board; for paneling, use trim screws. (Note: If the partition divides a heated from an unheated area, or must stop noise, saw the scribed 2xs to leave a ¼-inch gap between them and the irregular wall. Then, after permanently nailing them up, stuff ⅜-inch closed-cell backer rod into the ¼-inch gap. A good glass shop can supply the backer rod.)

Framing partitions

Finally, you're ready to nail some real framing wood. Begin with the double top plates, nailing them to either the ceiling joists or the nailers that you added as described above. If the old structure is so wavy that the plates won't fit tightly in places, don't hesitate to shim between the plates and the ceiling structure. Next, lay the sole plate in place, mark the location of door rough openings, cut out the openings, and nail the sole plate in place.

With plates in place, install the end studs where the partition joins older structure as discussed above. Install it tight to the existing structure, ignoring plumb. Then go to the door openings. Lay out both the trimmers and king studs on the sole plate, get out your plumb bob and mark the king stud locations on the top plates, measure the exact length required for each of the king studs, cut them to length, and toenail them in place.

Now lay out the common studs. Remember that the common studs are there primarily to hold up the wall covering, typically gypsum board, so think ahead about the most convenient way to arrange the wall covering. There may be quite a practical difference between working left-to-right and working right-to-left.

If the end that you intend to start with is seriously irregular or out of plumb, find the portion that is farthest out. Measure 16 inches in from this portion and plumb down to the sole plate. This is the center of the first common stud. Measure three quarters of an inch to either side of the center mark to find the edges of the stud. Mark the edges square across the sole plate, then plumb up to the top plates and mark the stud location on the top plate. Put a bold "X" between the stud marks to avoid confusion.

You could carry on, marking the locations of all the common studs, but if you're working alone you might want to cut and nail the first common stud in place so you'll have something to hook your tape measure on while measuring out the remaining studs. Just remember that all of the studs measure from that first one — don't measure in from both ends or change the spacing of common studs because of a doorway. Once laid out, cut each stud to correct length and toenail it in place. (*See pages 162 and 163 for help on toenailing.*)

CONSIDER AN ALTERNATIVE

It is not uncommon to see outbuildings renovated as studio apartments for older parents (or older children, for that matter) with the idea that the building can later be remodeled and put to other uses. You can make the remodeling job easier and more economical if you plan ahead. Consider installing nailers as shown in *Figure 7-4* where your partitions must join, then insulating, wiring, and installing gypsum board before adding the partitions. If you will need wiring in the partition, try to run it under the floor or through the ceiling rather than from the permanent wall into the temporary partition. When the permanent wall surface is in place, frame the partition using long drywall screws to fasten the end stud to the nailers inside the permanent wall. When the time comes to tear down the partition, you can do so with minimal disturbance to the permanent wall — a bit of joint compound and paint is all you'll need.

Nailers

Permanent walls

Temporary partition

Toenailing

There are three good reasons why you might want to use toenailing.

1. Obstructions prevent straight nailing.

2. Straight nailing would put a nail head where it isn't wanted.

3. Toenailing is stronger in some ways.

Many folks tend to shy away from toenailing because they find it frustrating and they don't trust it. That's unfortunate, because it need not be frustrating and it can be very trustworthy, indeed. First, the strength and reliability of toenailing in comparison with straight nailing needs a bit of explanation.

The holding power of a nail depends on both the direction of the nail and the direction of the forces bearing on the stock the nail is holding. When the nail is perpendicular to the forces against it, the nail must resist shear stresses. Nails are good at this. When the nail direction and the direction of the forces are parallel to each other, the nail must resist withdrawal. Nails with smooth, straight shanks are not very good at this. When the two directions are somewhere between perpendicular and parallel, the strength of the joint is also somewhere between.

In the typical stud wall situation our primary concern is sideways movement of the stud. Straight nailing through the plates and into the ends of the studs works well to prevent this movement. When obstructions require that we toenail, we can compensate for the lesser strength of individual nails by using more of them. This is what we do when toenailing studs between plates that are already in place.

Inexperienced framers are often uneasy because they don't know how much of a nail is actually going into the second piece of wood, or because they are afraid the nail will come out the other side before entering the second piece of wood. If this describes you, study the drawings below and get a good mental picture of the correct angles and positions.

Another source of uncertainty is deciding on what size nail to use. Experienced framers may use a wide variety of nail sizes for toenailing, everything from 16d (3½-inch) common nails to 8d (2½-inch) box nails. (Box nails are more slender than common nails of the same length.) For consistently good results, I recommend 10d (3-inch) box nails. They are long enough to do the job and slender enough to avoid splitting the stock.

The two most common frustrations when toenailing come from splitting the stock and from being unable to prevent the stock from moving off the marked position when starting the first couple of nails.

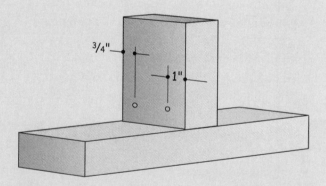

Place 10d box nails an inch or a bit less from the edges of the stud, two on each side, taking care to avoid hitting the nails from the other side. You can also nail from the edge of the stud but be sure to set the nail so it won't interfere with the wall surface material.

Choosing slender (box) nails is the first step toward preventing splitting. The second is careful selection and cutting of stock. Most construction lumber is kiln dried to about 19-percent moisture content. At this moisture content the wood is reasonably resistant to splitting by nailing. Lumber that has been stacked uncovered in the sun, however, quickly dries to a much lower moisture content, particularly at the ends, and becomes more prone to splitting. If you have much toenailing to do, keep the lumber out of the sun and try to use it as quickly as possible after it has been delivered. Also, select stock, or cut it, to avoid knots at the ends where you intend to toenail.

The ability to toenail a stud so that it stays exactly on the marks on the plates comes with experience, but it will come faster if you know a few tricks of the trade.

1. Pay attention to the angle of the nail. A steep angle has less of a tendency to move the stock. If the stock has already moved as you drove in the first nail, move it back with a nail from the opposite side. If necessary, drive the second nail at a lower angle to encourage it to move back.

2. Measure carefully and cut precisely. If you have to tap the stud lightly to get it in place it will be less likely to move as you nail it. Don't cut it so long that you have to drive it in with heavy blows, however. The forces you generate by driving a stud in place can easily raise up the entire ceiling. Several in a row will guarantee trouble.

3. If the stud slips too easily into place, put a shim of 15-pound roofing felt over the end as you put the stud in place.

4. Keep the stud from moving back by putting your knee up against the side of the stud opposite the nail.

5. If you're having trouble getting the hang of it, cut a scrap of 2x4 to just a little less than the space between studs (14½ inches) and place it against the previous stud to keep the current stud from moving.

6. Pay attention to how much a stud tends to move after the nail first enters the second piece and compensate by starting it that much further back.

My last tip is for framers who take great (excessive?) pride in their workmanship: Carry a mechanic's pin punch (roughly ¼ x 6 inches) in your tool belt and use it like a nail set for the last half inch or so when toenailing. It will give you nicely set toenails with nary an elephant track on the wood.

45°—55°

1"—1⅛"

Place the nail an inch or a bit more from the end, at the angle of 45° or a bit more, for the best holding power. To keep the stud from moving off the mark, start your nail hole a little higher (about 1¼ inches) and drive it at a steeper angle (about 70°). You can move a misaligned stud back on the mark by starting lower and driving the nail at a 30° angle.

Insulation, condensation, and ventilation

Insulating (and presumably heating) an outbuilding that was not originally intended to be heated requires some careful consideration. If not done properly the building can be very expensive to heat, and its life expectancy can be greatly reduced. The difficulties lie in two characteristics of many outbuildings: First, many outbuildings, such as hay barns and especially tobacco barns, were intentionally built to allow ventilation even while shedding rain. And, second, an efficiently heated building relies in part on steps taken during early construction, steps as far back as laying the foundation and erecting the main structure. To heat the building efficiently, you must not only insulate, but also stop excessive ventilation and prevent destructive condensation.

Condensation

Water vapor condenses into liquid water when humid air cools down. When this happens inside your walls in midwinter the way it does on your cool drink in midsummer, it greatly reduces the effectiveness of the insulation. If there is no way for the condensation to escape, the lumber in your walls will rot. To get an idea of how easily this can happen, *see Figure 7-7*.

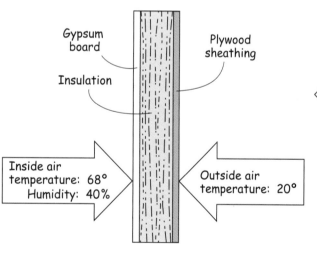

7-7. With walls of this type, conditions are ripe for condensation to start halfway through the wall and get worse from there to the outside. Water permeates gypsum board about 95 times faster than it permeates plywood, so a continuous buildup of condensation inside the wall is likely. Two coats of vapor barrier paint on the gypsum board would make the inner surface much more resistant to the passage of vapor. A layer of polyethylene sheeting under the gypsum board would also make the inner surface five times more resistant to the passage of water vapor than the plywood outer surface.

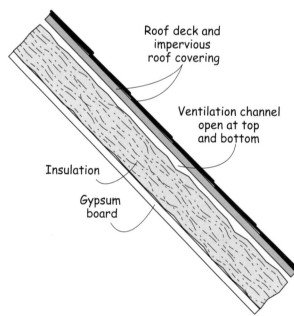

7-8. Insulated roofs are best protected from condensation problems by a ventilation channel between the roof deck and the insulation. The channel must be open to the atmosphere at the top, usually through a ridge vent, and also open at the bottom, usually through soffit vents.

How condensation works

Water vapor, the source of condensation, is the invisible gaseous form of water. It is normally mixed in with all the other gases that make up what we know as "air." When there is a lot of water vapor in the air we say that the air is humid; when there is little we say it is dry.

The steam that you see coming out of the tea kettle is not water vapor — it is tiny droplets of liquid water that have condensed out of the water vapor shooting out of the tea kettle spout. Clouds and fog consist of similar tiny droplets of liquid water.

Water barriers, vapor barriers, and air barriers are not the same. A water barrier stops liquid water but doesn't necessarily stop water vapor. An air barrier stops air from blowing through but doesn't necessarily stop water vapor. For example, a Gore-Tex rain jacket will stop wind and rain from coming through but still allow evaporated perspiration (vapor) to escape. Most materials that we use as vapor barriers, including aluminum foil and polyethylene sheeting, also stop water and air.

The best way to prevent problems from condensation is by making sure that water vapor has a harder time getting through the inner surface of walls than it does escaping to the atmosphere. If you do that, then any vapor that gets into the wall will do no harm. Engineering this great escape can get complicated and the technically minded are welcome to immerse themselves in condensation by reading the *American Society of Heating, Refrigerating and Air-Conditioning Engineers Handbook of Fundamentals.* The rest of us can simply look at typical solutions for typical situations.

If the outer surface of your walls consists of no more than the original, handsome, barn boards (typically softwood an inch or less in thickness), then vapor can easily escape. The joints between the boards will allow a great deal of air to ventilate the wall cavity and carry away vapor. Even ignoring this ventilation, the boards themselves are only one-seventh as resistant to the passage of vapor as plywood sheathing. Two coats of latex vapor barrier paint on the inner walls provides good protection.

If the outer surfaces of your walls have plywood on them, either plywood siding or plywood sheathing, then a more effective barrier is needed on the inner surface. Polyethylene sheeting, carefully installed with taped seams, will give good protection: Vapor will have five times as much trouble getting into the wall as getting out.

Ceilings with ventilated attic space above them seldom cause any trouble at all but cathedral ceilings definitely do. With cathedral ceilings the roof covering is often a very effective vapor barrier, trapping any vapor that gets past the inner surface. The only effective solution is a ventilated passage between the insulation and the roof deck. This is achieved by stapling a formed, expanded polystyrene ventilation channel to the underside of the roof deck between the rafters. It is essential that both the top and bottom of the channel be open to outside air, usually by means of soffit vents at the bottom and a ridge vent at the top.

EXTRA HELP

Make sure you have a workable plan for controlling water vapor and condensation before you begin renovating. If you plan to cross that bridge when you come to it you may come to wish — a little too late — you had renovated differently.

The catch when converting unheated outbuildings to heated buildings is how difficult it is to seal the inner vapor barrier up tight. A vapor barrier does no good if warm, humid air can find its way around the edges. For example, it is not uncommon to see a fine, old, timber-frame barn converted for use as a heated workshop or even as a dance or meeting hall, or to see a delightful old timber-frame carriage house converted for human habitation. Since the outside of the timber frame must be covered for its own protection, the owners usually want the frame exposed on the inside. The logical place to insulate is then between the posts, and that requires that the vapor barrier seal tightly to the posts on both sides of each bay, to the girt above, and to the sill below, and without showing in the finished interior. One way to go about this is shown in *Figure 7-9*. You'll read more about this approach in the section below under "Gypsum Board."

Ventilation

Ah, nice fresh air. Brrr, cold drafts. They're the same thing, differing only in quantity and in the season of the year. To get the benefits of fresh air without the cost of heating too much of it in the winter and without the discomfort of unheated drafts you need to control the air — it's that simple and that difficult.

Controlling ventilation comes down to stopping uncontrollable air leaks and providing controllable ones. The steps described above for eliminating condensation will go a long way toward stopping uncontrollable ventilation — air leaks around doors and windows and any other uninsulated parts of the building's exterior are all that remain. Caulk, weather-strip, tape, seal, plug — when renovating an old building just about anything goes. The difficulty is not so much finding a way to stop a specific leak as it is finding where all of them are. Go over the building inch by inch. Every edge of every element should be suspect until you've satisfied yourself that it's tight. When you spot a potential leak and plug it, ask yourself if there's a detour that the air could take and plug that too. You won't find them all, and a few that you do find you won't be able to plug satisfactorily. Just do your best and don't let the remaining ones cost you your night's sleep.

7-9. *For a tight seal with a wooden post, L-bead on edge of gypsum board is held ¼ inch back from post. A ⅜-inch closed-cell foam backer rod compressed into gap presses sheet against post. Trim off the poly sheet after painting.*

Providing controllable ventilation is the comparatively easy part. First, and most importantly, provide direct outside air to anything in the building that has a fire in it — furnace, boiler, wood stove, fireplace, and so on. If you don't, you'll get poor combustion and quite possibly dangerous fumes in the building. Next, provide positive ventilation to the areas that normally have it anyway, such as bathrooms, cooking areas, and paint rooms. These exhaust fans will suck in air through the leaks that you missed, through the ducts that you provided to combustion devices, and, if necessary, through a window opened a crack at the top.

7-10. Traditionally, barns and outbuildings were ventilated with simple openings in the walls. The decorative brickwork on this barn ventilated the interior.

If your heating system will be forced hot air, you have another, better, option for providing fresh air. Have your heating contractor run a small duct from the outside into the cold-air return duct. When the blower comes on it will suck air from the return air registers in the building and also through the duct from the outside. The mix will then be heated and distributed throughout the building. Stale air will exit through leaks in the house, through the one-way dampers on any local vent fans you may have installed. If necessary, you can install an additional vent with a one-way damper at the outside. The advantage of this system is that the inevitable leaks will be leaking stale air out rather than cold, drafty air in while the fresh but cold outside air is heated before it finds its way to your ankles.

Finally, if you're a fanatic for fresh air but hate to waste fuel, ask your heating contractor about air-to-air heat exchangers. These devices draw stale air from a building and use its heat to warm incoming fresh air. You get fresh air without having to throw away a lot of heat.

Insulation

The last few decades have witnessed an increasingly scientific or "engineering" approach to small building construction. One result of this trend has been a growing variety of suitable insulation materials and techniques. Consider your options carefully. Each material and technique has its own advantages and disadvantages.

Batts

Fiberglass batts are the most commonly used insulation material. They are economical compared to most alternatives and easily installed without special techniques or equipment, provided the framing of the building is on standard 16- or 24-inch centers. Most people who have installed fiberglass consider it to be a nasty, itchy-scratchy job, but one that goes fairly quickly. Batts lose some of their appeal when a building has a lot of small or irregular cavities that will require insulation. Unfortunately this is typical of many outbuilding renovation or conversion jobs. Keep in mind, too, that batts provide only insulation — they are ineffective as air or vapor barriers.

Loose fill

A variety of light, loose, more or less pourable materials have been used as insulation including tufts of fiberglass, expanded minerals such as vermiculite, and various forms of cellulose including plain sawdust. In the past these materials were most commonly used in attics where they were poured between the attic floor joists. As the economic advantages of insulating walls as well as ceilings became more apparent, inventive tradesmen devised ways of blowing these insulating materials between the plaster and the siding of older houses. Loose fill remains an economical and easily installed form of attic insulation.

High-tech installation techniques have also made cellulose fiber from recycled newspapers into an excellent form of wall insulation. The material is treated to make it fireproof and uninviting to insects and rodents. It is blown or sprayed into wall cavities before the inner surface of the wall is installed. The techniques used are extremely effective at filling irregular cavities, getting around behind electrical boxes, wiring, and plumbing, and generally accomplishing those things that are quite difficult to do with fiberglass batts. In some renovation jobs it's the most effective practical way to stop cold air from infiltrating the building. As an insulator, cellulose fiber is at least as effective as fiberglass batts, so its ability to effectively insulate voids that fiberglass batts can't get into makes the

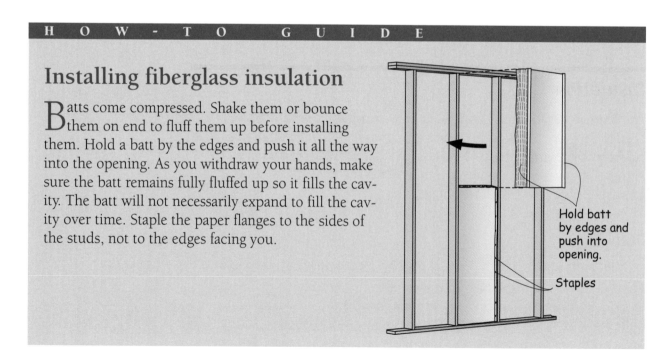

HOW-TO GUIDE

Installing fiberglass insulation

Batts come compressed. Shake them or bounce them on end to fluff them up before installing them. Hold a batt by the edges and push it all the way into the opening. As you withdraw your hands, make sure the batt remains fully fluffed up so it fills the cavity. The batt will not necessarily expand to fill the cavity over time. Staple the paper flanges to the sides of the studs, not to the edges facing you.

Hold batt by edges and push into opening.

Staples

overall insulation of a wall with blown or sprayed cellulose superior to fiberglass in most situations.

There's a catch, of course. It's not a do-it-yourself operation and it takes a substantial investment in equipment to get into the business. The result is that sprayed cellulose insulation is not commonly available outside urban areas. If you're reasonably close to a city and think this technique might just fit your needs, get out your yellow pages and start calling around.

Rigid foam

Several types of foam "boards" have come into common use as insulation in the last several decades. These include:

Expanded polystyrene, often called bead-board. This is the same material used for those white foam, throwaway coffee cups and for custom-shaped, white foam packing materials. As an insulator it is at least as good as fiberglass batts but should not be considered a vapor barrier.

Extruded polystyrene. This is the same kind of plastic but is made into boards by a different process. It does not have the bead-like structure

Characteristics of insulation materials

INSULATION MATERIAL	R-VALUE	VAPOR BARRIER	INFILTRATION BARRIER	COMMENTS
BATTS				
Fiberglass batts	3.0–3.8	no	no	Quick, simple if unpleasant to install, and economical for standard framing
LOOSE FILL				
Loose cellulose	2.8–3.7	no	no	Quick, easy, and economical between joists in attic
Vermiculite and Perlite	2.2–2.8	no	no	Quick, easy, and economical between joists in attic
Sprayed cellulose	3.0–4.0	no	yes	Excellent where available; requires equipped contractor
RIGID FOAM				
Expanded polystyrene	3.6–4.4	no	yes	Least expensive rigid foam; protect from water
Extruded polystyrene	5.0	yes	yes	Good choice for under concrete slabs and in roofs
Isocyanurate	5.6–6.3	yes	yes	Good choice for adding insulation under gypsum board

NOTE: The R-value expresses the insulating effectiveness of a material. The number generally increases with the thickness of the material, so doubling the thickness usually doubles the value. Insulating practices vary with climates, local codes, and budgets, but inhabited spaces typically have R-19 or higher insulation in the walls and at least twice that in the roof.

of expanded polystyrene. The most common brands are colored blue and pink respectively. As an insulator it is about 50 percent better than an equal thickness of fiberglass batt, plus it is fairly good as a vapor barrier.

Urethane and isocyanurate. These insulation boards are nearly twice as good as fiberglass batts, inch for inch. Since they are typically faced with aluminum foil, they are excellent as vapor barriers. The aluminum foil also makes them helpful at stopping radiant heat loss.

While all of these insulation boards have a legitimate place in old building restoration, that place is rarely as a substitute for batts, loose fill, or sprayed cellulose. Their higher price for a given degree of insulation dictates that they be used only when they offer other advantages such as compressive strength or the ability to double as a vapor barrier. Both of these advantages can come into play if you're renovating an outbuilding framed with 2x4s but want more insulation than three to four inches of fiberglass. In this case you could use fiberglass between the studs and then a layer of rigid foam over the studs, under the gypsum board. The compressive strength of the foam allows you to screw the gypsum board through the foam to the studs without crushing the insulation board. If you've chosen extruded polystyrene or isocyanurate foam and carefully taped the joints, you've simultaneously installed an effective vapor barrier. You've also saved yourself the time and expense of adding all new framing just to get more room for fiberglass.

Another common use for rigid foam insulation is in roofs. Say, for example, you want to leave the underside of the roof deck boards and the rafters or purlins exposed as your finished ceiling but want to add insulation. You can nail the rigid foam to the rafters or purlins right through the deck boards, then nail a layer of plywood or oriented strand board over the foam, and finally nail the finish roofing to the plywood or oriented-strand board (OSB). This procedure won't come close to the usually recommended amount of insulation for dwellings in cold climates but could well be adequate in warmer climates or for structures that don't need to be heated to human comfort levels.

Gypsum wallboard types and thicknesses

TYPE	APPLICATIONS
Regular	Most applications not subject to unusual abuse.
Water resistant	Walls and ceilings subject to high humidity, steam, or splashed water.
Fire rated	When required by local codes; where subject to greater physical abuse, such as garages, utility rooms, and shops

THICKNESSES	
½ inch	Walls framed not more than 16 inches on center.
⅝ inch	Walls framed not more than 24 inches on center.

Installing gypsum wallboard

Gypsum wallboard, commonly called drywall, has almost entirely replaced lath and plaster as the interior wall surface material of choice. Briefly, gypsum wallboard consists of large sheets of a gypsum-based material between sheets of heavy paper. Joints between sheets are concealed with a plaster-like material called joint compound reinforced with an embedded tape, typically of paper. The drywall is "hung" and then "taped."

Compared to plaster, drywall is faster and requires less skill to install, and is therefore less expensive. It is also less expensive, more fire-proof, and a better acoustic barrier than wood paneling. These advantages have made it so common that many people want something else just to be different. Be different if you want to — just be aware of the advantages of drywall that you are giving up.

While drywall installation does not require great skill, it does require some knowledge and experience to get an installation that will look good and remain free of cracks.

One of the more important considerations is the layout of the sheets. Drywall comes in sheets 4 feet wide and from 8 to 16 feet long. The long edges are tapered as shown in *Figure 7-11* so they can be taped to give a seamless appearance. The 4-foot edges are cut square, are more difficult and time-consuming to tape, and are seldom completely invisible. Clearly, the best job minimizes the joints between square-cut edges, which are known as butt joints.

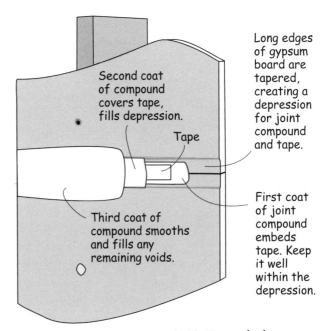

7-11. Tapered edges, embedded tape, and three coats of compound create an invisible seam.

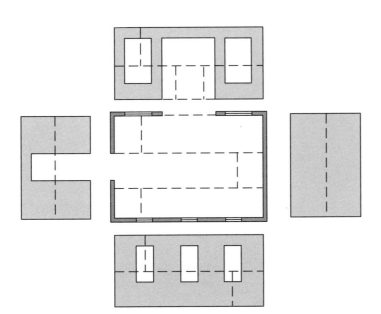

7-12. This small outbuilding measures 17 feet by 11 feet. Gypsum board comes in lengths of 8, 10, 12, and 14 feet. The best configuration of boards to use limits the number of butt joints, leaves enough room to feather out any necessary taped butt joints, and places them as inconspicuously as possible. In this case 14-foot sheets work out well. For the ceiling, you could instead use 12-foot sheets cut to 11 feet and installed without butt joints, but it is easier to hang 14-foot sheets perpendicular to the ceiling joists.

Nevertheless, there are situations where an extra butt joint or two is the wiser layout. Again, the consideration is taping. It is much faster and easier to get a good taping job if the entire joint is within easy reach and you can sweep the trowel from one end of the joint to the other in one smooth, continuous stroke. That's a tall order if the joint is vertical, from floor to ceiling, or runs up the slope of a cathedral ceiling. Experienced drywallers therefore hang walls and cathedral ceilings horizontally, even if that requires butt joints.

The cracks that appear after a few months or a year in some drywall jobs usually occur at corners. Where walls meet adjoining walls or where walls meet ceilings, cracks can be minimized by using "floating" corners, as shown in *Figure 7-13*. At the corners of windows and doors, the best ways to prevent cracks is to avoid joints between sheets at the corners, as shown in *Figure 7-12*. This practice also makes trimming out the doors and windows easier because there is no buildup of joint compound where you want the casing to lie flat. In practice, this means hanging the gypsum board right over door and window openings and then cutting them out with a drywall saw or small router fitted with a special drywall cutout bit.

One of the problems frequently encountered when restoring old buildings is finding an attractive way to treat the juncture between modern materials like gypsum board and old building elements like hand-hewn posts and beams. The most common solution is to apply 3-inch-wide masking tape to the post or beam before installing the gypsum board, then apply joint compound right up to the hand-hewn irregularities. The tape is left in place until after final painting, then cut with a razor blade and removed.

The masking tape routine works, and is preferred by many people, but it never looks right to me. The irregular edge on the flat, smooth, painted wallboard doesn't seem appropriate. I prefer to keep the edge of the gypsum board straight and leave a small gap between it and the old, irregular beams. The gap allows each element to retain its own integrity. Be forewarned, however, that keeping the edge dead-straight and retaining that gap is a bit of a time-consuming trick. I use the procedure described on page 166 in the section on installing vapor barriers.

The drawings in *Figures 7-14* and *7-15* on page 174 demonstrate the final steps in installing wallboard this way. Here are the steps to get there.

Hang the gypsum board leaving a straight-edged ¼-inch gap between it and the post or beam. If you have a vapor barrier behind the gypsum board, make sure you don't bury the edge when the gypsum board goes up.

Apply metal edge-bead, commonly called L-bead. If you're treating a cutout for a ceiling beam that doesn't have an exposed post below

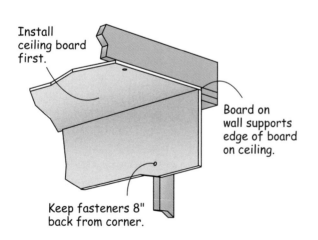

Install ceiling board first.

Board on wall supports edge of board on ceiling.

Keep fasteners 8" back from corner.

7-13. *This floating corner procedure is sometimes also used where walls meet walls.*

When the situation requires butt joints in gypsum board, the conventional method is to join the two sheets on a framing member, a stud, or a ceiling joist. But even if the framing member is perfectly placed behind the joint, the gypsum tends to crack and crumble when a nail or screw goes in too close to its edge. The job is even more difficult in an old outbuilding where the framing is often not parallel or plumb, not on uniform centers, or is twisted.

An excellent alternative is to position a butt joint more or less midway between framing members, with a board or strip of plywood backing up the joint. The procedure is simple: Place or cut the first sheet so it ends between framing members and screw it in place. Clamp or hold the backup board or plywood strip halfway behind the edge to be joined and screw through the gypsum board into the backup. Then hang the joining sheet, screwing its edge into the backup at the butt joint.

Note: This alternative is practical only if you are hanging the drywall with screws, using a screw gun. If you are nailing the gypsum board, you will have to install an additional framing member behind the joint.

Softwood lumber an inch or so thick and about 4 inches wide makes a good backup provided it is not too knotty or prone to splitting. Leftover strips of plywood sheathing also work well — screws hold very well in it, although they are a bit more difficult to start.

TO BE SAFE

If you are holding the backup by hand while screwing into it, be extremely careful not to screw into your hand — hefty spring clamps are a lot safer.

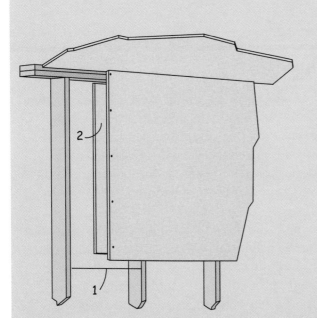

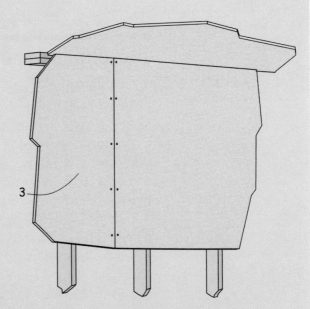

1 When framing is out of plumb or not on uniform centers, make drywall butt joints between framing members instead of on them. Hang the first sheet so it ends between framing members.

2 Clamp and then screw a backup board behind the proposed joint.

3 Hang the second sheet, screwing the end to the backup board. Join sheets of the next course (the sheets above or below) in a different bay.

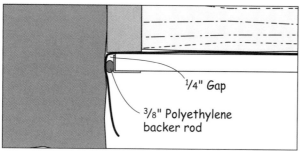

7-14. *This method provides a small, straight gap between gypsum board and a beam.*

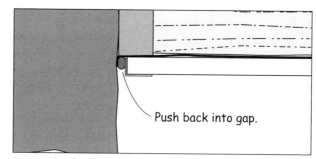

7-15. *The final result leaves the backer rod discreetly tucked in the gap.*

it, use a single piece of edge-bead by cutting through the face flange and folding the edge flange to create a U shape.

Now insert a ⅜-inch closed-cell polyethylene backer rod into the ¼-inch gap but don't push it all the way back into the gap; leave it projecting out to keep joint compound from filling the gap.

Embed the edge-bead in joint compound just as you would a corner bead or taped joint. The backer rod will keep compound out of the gap and the protruding vapor barrier will protect the post or beam. After painting the wall, break off any joint compound that stuck to the backer rod and push the rod as deep as it will go into the gap. Trim off the excess vapor barrier.

Installing paneling

Many different wall surfaces are called "paneling" these days, everything from old barn boards to elaborate raised panels set in grooved frames — an entire book devoted to wall paneling would hardly exhaust the subject. The modern manufactured materials available at home centers typically have available accessories for treating edges, corners, baseboards, and ceiling junctures as well as guides for installation, so no more needs to be said here.

At its high end, paneling is the province of skilled and experienced woodworkers who, for a price, will install wall coverings that differ little in quality or workmanship from fine furniture. More frequently, when "paneling" is used in restored outbuildings it is in the form of boards, often with tongues and grooves along the edges and sometimes with molded shapes on the face along the edges. This face molding can be a simple V groove or "bead" that helps disguise the tongue-and-groove joints, or it can be a more elaborately molded band an inch or two wide, such as was common on knotty pine paneling in the 1950s. Square-edged boards are also used as paneling. This category includes the weathered "barn boards" that rise and fall in popularity — some people love them so much they want them on the walls of their rooms; others love them only when they're still on the outside of the barn.

Perhaps the biggest secret to doing an attractive job of installing board paneling is paying attention to the details — not only how to do the job but also how it will appear when it's done. Spend some time looking carefully at how other installations were done, and consider which of their details you admire. A few general suggestions follow:

Provide adequate backers or nailers between the studs. Most board paneling is installed vertically, so the studs themselves won't give you something to fasten to where you need it. If the boards are ¾ inch or more in thickness, you will need nailers every 32 to 36 inches, including both top and bottom.

Think about what you plan to do where the wall meets the floor. If the boards have face molding and you want a baseboard, consider butting the paneling boards to the top of the baseboard. This avoids an unattractive gap where the molding would otherwise disappear behind a baseboard. You can do the same at the top, with or without a crown molding where the wall meets the ceiling.

If you're using weathered barn boards, consider a recessed baseboard. Leave the weathered bottom end of the boards intact, overlapping the baseboard and ending 6 inches or so above the floor. This comes closest to mimicking the appearance of barn boards in their "natural habitat."

Choose your fasteners to match the character of the boards. If the boards are nicely molded and sanded and will be finished with a stain and clear finish you might blind-nail the boards as shown in *Figure 4-33* on page 90. If the boards are more rustic consider face nailing them with cut nails.

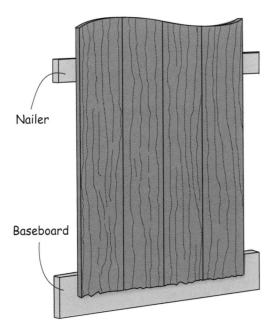

Nailer

Baseboard

7-16. *When paneling with barn board, consider a recessed baseboard in order to preserve the weathered bottom end of the boards. Install the baseboard first, then nailers of the same thickness. Face nail the boards with cut nails.*

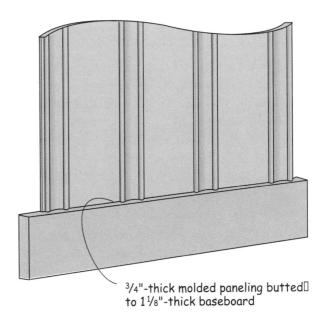

¾"-thick molded paneling butted to 1⅛"-thick baseboard

7-17. *Consider butting ¾-inch-thick molded paneling to the top of a baseboard 1⅛ inches thick in order to avoid gaps where the molding would disappear behind a surface-mounted baseboard.*

Planning partitions

Partitions can do a lot to make spaces more useful and efficient, but if positioned poorly they can also make spaces much less usable. Careful planning is essential. If you're not especially good at visualizing the effect of a partition in a particular location, don't just hope it will work out well — try it out by putting up big pieces of cardboard or hanging a drop cloth or a sheet. Then judge the partition by the considerations that follow and maybe even live with your temporary divider for a while. You could save yourself a lot of grief later on.

Partitions can provide privacy and an intimate ambiance. On the other hand, they may destroy a feeling of openness and spaciousness that is often prized in barn conversions.

They can isolate noises, odors, and visible clutter. But they also isolate occupants and prevent easy communication with people on the other side on the wall.

They can confine heated and cooled air to a smaller area, possibly reducing utility costs, though in doing so they may leave other spaces too hot or cold.

An arrangement of partitions that fully encloses a space needs an opening for a doorway and usually also requires a door in the doorway. But the swing of a door seriously limits the use of a substantial amount of space. Pocket or sliding doors intrude much less into other spaces but have their own disadvantages. They are less convenient to use and are usually less effective at confining noise and speech, odors, and heated or cooled air.

Give careful thought to traffic flow and furniture placement when planning partitions. A space intended as a conversation area or place to sit and read can be disappointing if the overall room arrangement sends traffic right nearby. On the other hand, a space can serve more than one purpose if its different functions seldom occur at the same time. For example, the dressing area of a bedroom might double as the pathway from the laundry to an outdoor clothesline.

A conversation area can be a bit too big or too small. Culture, climate, even personality affect the conversation distance that people prefer. Plan your partitions so that your furniture will fit the room while maintaining the proper amount of space to make people in the room comfortable.

Finally, when placing partitions and rooms, consider how they will affect your use of heating, lighting, and utilities. For example, closets and other storage areas can substantially reduce heat loss during the winter if they are arranged along exterior walls, especially those facing north. One arrangement of partitions may create a number of dark spaces you will need to light, while another may allow more bright southern light to continue to stream in.

I've used examples from residential arrangements here because everybody should be able to relate to them. The drawing on the opposite page provides a possible room and partion arrangement for a small barn or carriage house converted to a guest house or perhaps a retirement home. Consider the same effects of partitions when restoring an outbuilding for other purposes, including office use, shop use, garage use or just about any use that might demand partitions.

Clothes closets and coat closet help insulate north wall.

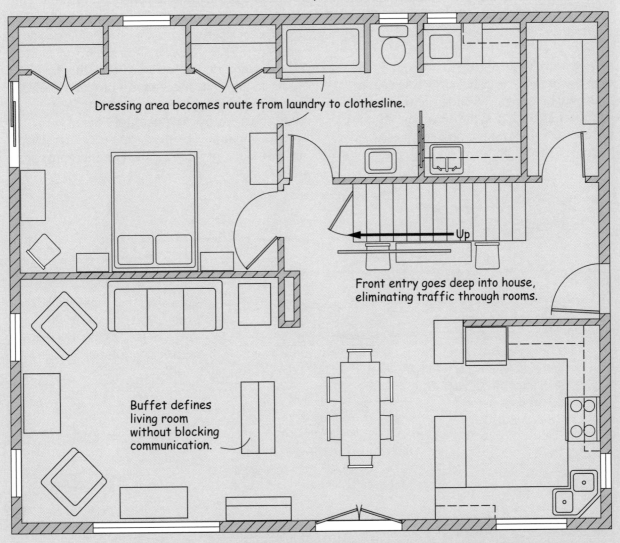

Dressing area becomes route from laundry to clothesline.

Front entry goes deep into house, eliminating traffic through rooms.

Buffet defines living room without blocking communication.

Up

Position doors for minimum intrusion into room space.

People who have done little construction work sometimes look down on the use of shims as a kind of cheating or as a cheap fix for a mistake. While shims can often make up for minor gaps and misfits, that is not their primary task in a building. Their real jobs are, first, to make fine adjustments in the position of assemblies like doorframes and windows and, second, to bridge between rough structural elements and more precise finish elements. In some cases, they do both — shims are used to bridge between the structural or "rough" opening and the door or window frame while at the same time allowing the door or window to be properly positioned with great precision.

When restoring an old building, shims have yet another function. They compensate for the irregularities that result from age and to some extent from the hand techniques used to shape older buildings in the first place. Even if a main floor beam in a barn were perfectly straight when the barn was built, it would be surprising if it did not take on a permanent bow after holding up tons of hay, year after year, for a century or more. When you try to fit a partition below that beam, however, it may be much more efficient to shim a straight partition to a bowed beam than to build a matching bowed partition. Get to know various shimming techniques to do a better job, not just to fix mistakes or find an easier route.

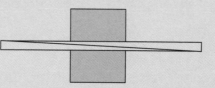

The most widely recognized and used shimming materials are cedar shingles. Used in pairs, they can be very precisely adjusted. Shim shingles are typically $^7/_{16}$ inch thick on one end, $^1/_{16}$ inch thick at the other end, and 16 to 18 inches long.

Paired, as shown above, they are about $^1/_2$ inch thick at the mid-point. Pushed all the way in, they shim a space about $^{13}/_{16}$ inch thick, and pulled back they shim a bit over $^1/_8$ inch.

Each shim that is unbalanced by another in the opposition direction fills a space with sides a bit over one degree out of parallel.

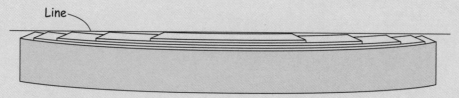

Line

Roofing felt and similar products are often cut into strips and stapled into concavities to shim them up to a flat surface. Thirty-pound roofing felt is about $^1/_{20}$ inch in thickness. The more common 15-pound felt is thinner, and granulated roll roofing a bit thicker. Self-adhesive bituminous membranes are too expensive to buy just as shims, but scraps can be extremely handy where it is impossible to staple a shim in place.

Build up layers to a straight line as shown, pounding staples down firmly to prevent them from creating a false height.

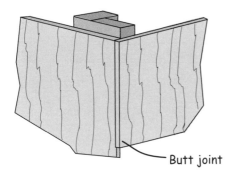

Butt joint

7-18. When installing boards with an irreproducible finish like aged barn boards, trim the board at left first to a straight edge and install it, then start the second wall with the board at right, leaving the original edge showing and covering up the trimmed edge of the first board. For recently milled boards, install the left board, then install the right board slightly proud of the adjoining surface. Use both glue and nails for the joint. When the glue is dry, sand or plane the joint flush.

Conceal any cut edges of the pieces in outside corners. With some materials, a newly cut edge won't match the rest of the board. It's usually best to lap a board with an uncut edge over the cut edge of the last board on the other side of the corner. If you're using newly milled stock, glue as well as nail the lap joint and then plane it flush after the glue dries.

If your boards do not have tongues and grooves, consider what you will see through the inevitable cracks between them. Remember that those cracks will be wider during the dry winter months than they will be during the humid summer. You may want to install vertical battens behind each joint so that natural wood will appear behind the cracks. Another alternative is plain, old, black roofing felt stapled to the wall before installing the boards. Dull, flat, black at the bottom of the crack can hardly be seen at all. The worst job I ever saw showed the aluminum facing of isocyanurate foam insulation between the boards. Time spent thinking about the end result will be well repaid.

Fitting walls in post-and-beam structures

Interior walls can be very important in dividing large spaces into rooms and providing places for utilities, doorways, or just to hang a picture. That's why you're reading this chapter. But one of the main attractions of an older barn or outbuilding is often its timeworn, handcrafted beams. That may be one of the reasons why you decided to restore your barn or outbuilding in the first place.

When locating an interior partition inside a bent (a part of the post-and-beam framework), you will often be able to construct the wall so that the beams will stand slightly proud of the wall on both sides. A wall constructed of 2x4s with a sheet of drywall on each side (which would be the normal design of an interior, non–load-bearing wall) will be about 4½ inches thick, or

7-19. Rough-hewn beams are attractive, but incorporating walls into their uneven corners and edges can be difficult.

slightly more. Such walls can be centered within the beams of a bent and will leave about 1½ inches showing on each side of a 8x8 post. You can form the transition from drywall to beam as described at the bottom of page 173 and shown in *Figures 7-14* and *7-15* on page 174. A simpler (if less attractive) way to meet the wallboard to the beam is to use a strip of L-shaped metal taping angle. Push the taping angle into the corner formed by the timber beam and the 2x4 stud you've nailed into it. Attach the corner angle to the stud. You can then put up the drywall on your stud wall and tape the end of the wallboard to the corner angle.

When framing walls within post-and-beam constructions, you will often need to work around large, irregular beams and the braces that connect the vertical posts with the beams above them. This may require some creative construction of the 2x4 framing that is the basis of your interior walls. When fitting drywall into these unusually shaped frameworks, you may want to use the technique outlined in the How-To Guide below for cutting pieces of gypsum board to fit.

Fitting drywall

If you want to allow the beauty of rustic, unfinished beams to show through while still covering your interior walls with drywall, you're going to end up with a lot of oddly shaped nooks and crannies. Here's an easy way to measure and cut drywall pieces of the proper size.

Cut thin wooden strips in a variety of lengths, and tack them together along the edges of beams that you are fitting drywall to. Carefully transfer the shapes you've created to a sheet of drywall laid flat. If the tacks extend just slightly through the strips, you'll be able to use them to hold the template in place on the sheet of drywall. Close to the inside edge of the template, score the drywall with a utility knife or other drywall cutting tool, remove the template, and cut out the shape.

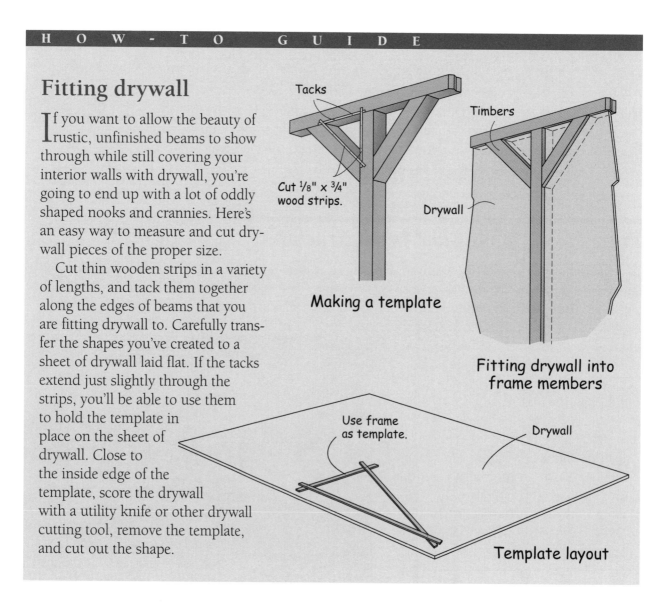

Making a template

Tacks

Cut ⅛" x ¾" wood strips.

Timbers

Drywall

Fitting drywall into frame members

Use frame as template.

Drywall

Template layout

Fitting walls in pole barns

Your barn or outbuilding may be built with pole construction, where the structure is supported by round poles set deep into the ground (this technique is sometimes used in order to avoid having to construct a full foundation). Poles may be situated outside the exterior walls, in the middle of the exterior walls, or entirely within the barn itself.

If the poles are outside the barn walls, and the exterior walls are of ordinary frame construction, then you can finish the exterior walls and attach partition walls to them the same as you would in other buildings with frame construction.

In many barns the poles will be inside the building. This design makes it possible to construct a larger barn or outbuilding with the same number of poles spaced the same distance apart. (The barn's structure can extend outside the supporting poles.) In those situations, you can build walls and partitions flush with the pole as in *Figure 7-22* or with the walls proud as shown in *Figure 7-23*. With poles that are into the interior of the space, you will likely have to drywall both sides of the frame.

One important consideration when fitting walls to any poles inside a barn is whether you want the poles to be exposed or completely concealed. The poles themselves may be unattractive, or they may be treated with wood preservatives that give off an unpleasant odor or are toxic. This is particularly likely to be true in older barns.

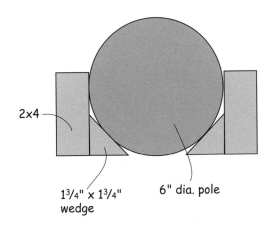

2x4

1³/₄" x 1³/₄" wedge

6" dia. pole

7-22. *Attaching frame to pole structure with wall flush.*

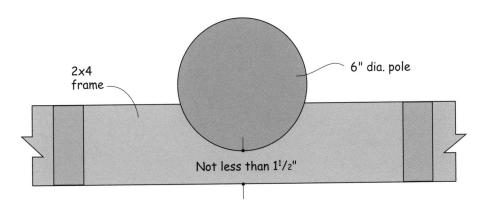

2x4 frame

6" dia. pole

Not less than 1¹/₂"

7-23. *Attaching frame to pole structure with walls proud.*

Converting a barn to a living space

One way to extend the long life of a barn is to convert into a home. In areas where farmland has been eaten up by commercial development and housing, conversion into a residence may the only way to enable an owner to preserve a large, historic structure similar to the barn depicted below.

In the How-To Guide on pages 176 and 177, I discussed the necessary considerations and options when you are converting a small barn into a guest house or in-law cottage. All of the suggestions there for determining room sizes, imagining traffic flow, and working to decrease utility use would also apply to the project of converting a larger barn into a full-size house.

Utility use is an even bigger concern in a larger barn conversion. The first and most important step is to install the maximum degree of insulation in the exterior walls that you can. Barn conversions often have open, soaring interior spaces; these make prevent-ing heat loss from inside the home an even more difficult task. Maximizing window openings on southern exposures and placing closet and storage spaces on northern walls can help bring down your heating bill.

One of the few signs that this home led a former life as a barn is the somewhat unusual window arrangement.

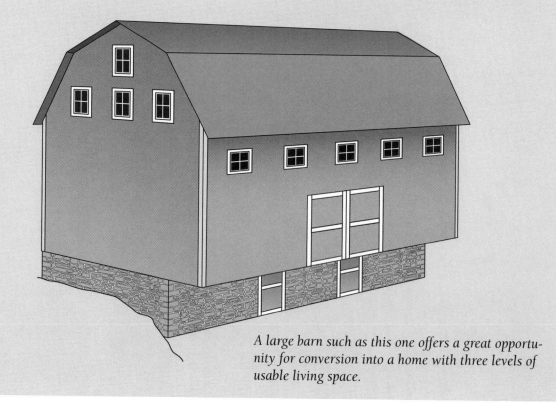

A large barn such as this one offers a great opportunity for conversion into a home with three levels of usable living space.

In the suggested floor plans at right, an open stairwell through all three floors helps to spread light and spaciousness to the new barn home. Another option is to leave certain living spaces open a level above their own. The bedroom and bathroom shown on the left side of the second floor could, with the elimination of the bedroom above them, become a dramatic master bedroom suite with open space up to the rafters of the roof.

In the floor plans here, the bathrooms and kitchen are all in the lower left corner of the house. Installing plumbing to run through those rooms and directly to the basement below where a hot-water heater might be located would be easier and less expensive than running pipes across the house.

Some of the rooms labeled as bedrooms on these plans could be instead used for storage. The third floor has storage along both long walls. In a barn with a gambrel roof such as the one in the drawing on the facing page, those areas may have short but still usable space. A barn with a more steeply pitched roof may have even less space that you can reach for storage on its upper level. Barns are frequently the spots where we toss all sorts of things that we don't know where else to store (and don't know enough to throw out), so converting one into a home can take away a lot of storage space itself.

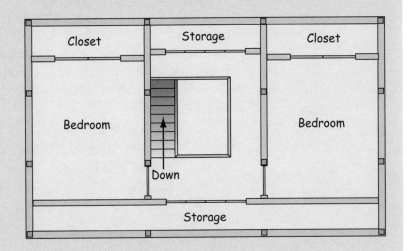

Third floor

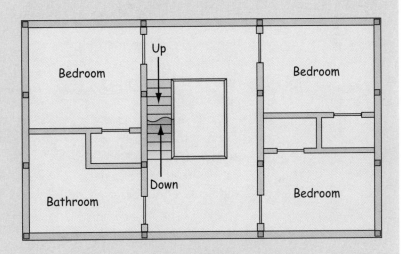

Second floor

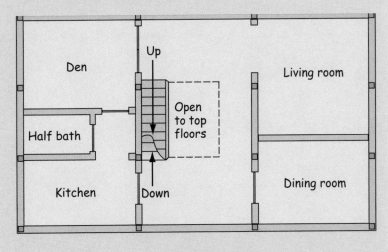

First floor

Attaching to masonry

Another special circumstance you may encounter is attaching an interior wall to masonry, whether it is of brick or block construction. This situation could occur in a brick outbuilding, of course, or you might want to put a wall in the lowest level of a barn or outbuilding with a masonry foundation.

Your first step will be to attach a 2x4 post to the masonry wall in order to have a support for the rest of the stud frame wall. As shown in *Figure 7-24*, you will need to drill holes into the mortar between the bricks or blocks of the proper size to accept a lead anchor. Use a plumb bob or level when marking where to drill to make sure that the holes are in a perfectly vertical line. Countersink a lag screw through the 2x4 (so its head won't conflict with the wall you're about to add), and continue to construct the wall as described earlier in this chapter.

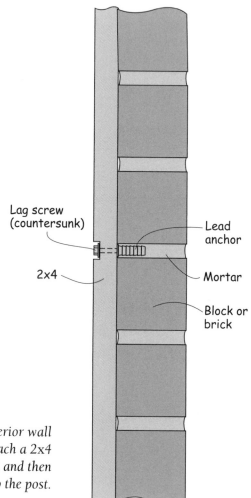

Lag screw (countersunk)

2x4

Lead anchor

Mortar

Block or brick

7-24. To connect an interior wall to masonry, first attach a 2x4 post to the brick wall and then connect the new wall to the post.

Roofing

The roof is the single most important part of the exterior of a building. It is the primary provider of shelter for both the contents of the building and the structure of the building itself. Those picturesque New England covered bridges weren't built that way for their quaintness — they were roofed because the bridge's structure was too valuable to be left to rot in the elements. Even if you have no current use for an outbuilding that you own, keep its roof in good repair. If you don't you'll soon have the considerable expense of having it torn down and hauled away.

185

Restoring an outbuilding roof can mean anything from fixing a leak to replacing the entire roof. If you do end up replacing the entire roof, keep in mind in the future that fixing roofing problems as they arise is the only way to prevent having to replace the roof all over again.

Fixing roofing problems

For some reason, when people go to repair a roof they often forget everything they know about how roofs work. The roof leaks, and they climb up with a bucket of roofing cement to "plug the leak" as though the roof were a continuous waterproof membrane. But the roofs of barns and outbuildings are usually pitched, like a hill, and covered with many more or less waterproof parts arranged so that when water flows to the edge of one part it drops down onto another part, and so continues until it reaches the eave. The most obvious parts used in this way are asphalt or wood shingles,

Getting to the job site

The first task in any roofing job is getting up there; the second is staying up there until you decide you're ready to leave. If you safely manage both of these you can look upon inconveniences like cuts, scrapes, and slivers in proper perspective.

The key tool is, of course, the ladder, but not just for getting up high enough to peer over the eave. Equipped with a hook, the ladder also gets you up to the ridge. And when two ladders are hooked over the ridge and equipped with scaffold brackets, they can support scaffold planks that allow you to traverse the roof.

You don't need a ladder to do roof work — you need several. And they need to be in top-notch condition, as well maintained as an airplane. The main difference between a fall from the ridge of a big barn and a fall

from a 30,000-foot altitude is how long you have to think about it on the way down.

If you're not already familiar with ladders on roofs, it's a very good idea to work with someone who is. Many of the steps involved, such as carrying up scaffold planks, are much more safely done by two people.

One final and essential tip: Keep an eye on your fear level. If you find that you're totally fearless, get down — you're a danger to yourself. Roof work is dangerous and a total lack of fear is unrealistic. On the other hand, you're also a danger to yourself if you're shaking with fear. Work between the two extremes, preferably closer to the bottom of the fear level but not at it. You'll probably find as you spend more time on the roof that your comfort level will increase.

roofing slate, and roof tiles. Less obvious, but also used for the same goal of guiding water away from where it isn't wanted, are flashing, ridge caps, and many forms of metal roofing.

When a roof leaks, it is usually for one of two reasons:

1. One or more of the generally waterproof parts is no longer waterproof. A slathering of roof cement may keep out the rain for a few months, but it's not a repair. To repair the roof, replace the deteriorated part.

2. One or more of the parts — a shingle, a slate, a piece of flashing — has shifted out of place so that it no longer channels water to the top of the next part. To repair the roof, secure the parts, or their replacements, in the proper position.

Replacing shingles and slates

Before you get yourself all geared up to replace shingles or slates, you should ask yourself what caused the problem in the first place. If the problem resulted from a falling tree limb, or wind that tore up just one small area, replacement of the affected parts may make sense. However, if the cause of the problem is deterioration of the roof's materials then you need to examine the rest of the roof. It doesn't make sense to repair just a portion of the roof this year if you're going to have to repair or replace the whole roof within a couple of years anyway — you'll be better off doing the whole job now. Consider also that your clambering about on a roof is

8-1. With many barn roof repair jobs, simply getting up to the roof is the most difficult part of the job.

8-2. The essential task in installing shingles is to channel water to the edge of the roof.

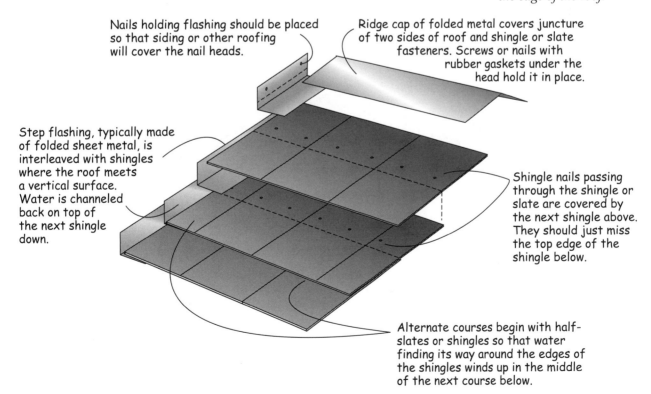

Nails holding flashing should be placed so that siding or other roofing will cover the nail heads.

Ridge cap of folded metal covers juncture of two sides of roof and shingle or slate fasteners. Screws or nails with rubber gaskets under the head hold it in place.

Step flashing, typically made of folded sheet metal, is interleaved with shingles where the roof meets a vertical surface. Water is channeled back on top of the next shingle down.

Shingle nails passing through the shingle or slate are covered by the next shingle above. They should just miss the top edge of the shingle below.

Alternate courses begin with half-slates or shingles so that water finding its way around the edges of the shingles winds up in the middle of the next course below.

8-3. With slate roofs, the problem is sometimes with the fasteners, not the slate itself.

often the greatest abuse the roof ever gets. If the slates or shingles are on the brink of falling apart it will be almost impossible to replace those that have already deteriorated without destroying more than you have replaced.

Asphalt shingles lose their flexibility and become brittle as they age. To replace shingles you need to be able to bend back the lower portion of the shingles above, remove the nails holding the shingles to be replaced, remove the bad shingles, slide replacements up under the shingles above, nail the replacements, and bend the upper shingles back into place. Go at it cautiously until you're confident that the shingles still have sufficient flexibility. If they don't, start thinking "new roof."

Slate roofs experience two kinds of deterioration: the slates themselves and their fasteners. Good slate is quite impervious to moisture. Lesser slate can absorb moisture. If it then freezes, the expanding moisture weakens the slate. After many years it begins to crumble. Good slate that hasn't been subject to physical abuse will last for centuries but its fasteners may corrode.

8-4. Nail holes are punched from the back of the slate toward the front with a slater's hammer, which has a pick opposite the head (above right). As the pick breaks through the front, flakes of slate break off leaving a depression for the nail head. When slates are hung on hooks (below right), the hooks are placed between the slates in the last course, just above the slates below. The long portion of the hook should not project above the surface of the slates but fit down between them. Standard hooks are 3 inches long, the same as the usual headlap.

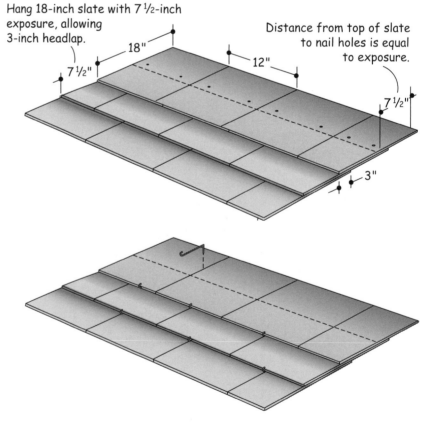

Hang 18-inch slate with 7 1/2-inch exposure, allowing 3-inch headlap.

Distance from top of slate to nail holes is equal to exposure.

Slate is "hung" with either nails or hooks. When hung with hooks, the lower hook portion is exposed to the elements and eventually rusts. With the hook gone, the slate slides out.

Replacing slates hung on hooks is fairly straightforward. If the slate is still in place, bend the hook straight and pull out the slate. Then replace the old hook with a new one, slide in the new slate to just beyond the hook, and pull it back into the hook.

Replacing nailed slate requires a special tool called a slate ripper that slides up under a slate, hooks around a nail, and then can be hammered to cut off or pull out the nail. If you have a slate roof you are probably in an area where slate roofs are reasonably common. If you ask around you will probably be able to borrow this indispensable tool. Having pulled the nails, pull out the slate and replace it using a slate hook as described above.

Replacing parts of the deck

First, check that only part of the deck is in need of replacement. You don't want to get halfway into this operation only to find out that the job is five times as great as you thought. In many barns and outbuildings the place to check the deck is from its underside. Poke and prod all over with a sharp awl, searching for soft spots. Once you're satisfied that only a local repair is needed, gather up your tools and check the weather forecast.

Replacing part of a deck may be time-consuming but it's not complicated. Starting at the top of the area that needs replacement, carefully remove the roofing as though you were replacing shingles or slates. With the top row removed, it's much easier to remove the roofing all the way down to the bottom of the area that needs replacement. Make sure that you remove roofing to a foot or so beyond the nearest rafters unaffected by the deck problem. Now remove the affected deck back to the nearest rafters. Install new decking of the same thickness, cover it with roofing felt, and re-shingle or re-slate.

8-5. When replacing damaged shingles, check to be sure deterioration has not reached the deck below them.

Replacing flashing and valley liners

Flashing and valley liners are normally folded sheet metal — copper, galvanized steel, or aluminum. Though they typically last as long as the roof covering, and much longer if the covering is asphalt shingles, they nevertheless corrode away in time, can be blown out by unusually strong winds, or may be torn from their fasteners as a building shifts and settles over the years. If your renovation has included straightening up the building, don't be surprised if your flashing begins to leak — it may have been installed after the building had settled.

The basic procedures for replacing flashing and valley liners are just like the procedures for replacing parts of the deck: Remove the shingles or slates that overlap the edges of the metal parts, remove the parts that you need to replace, fit the replacements, and re-shingle or re-slate.

If the problem is a leak that seems to originate in a valley but you're unable to find an obvious flaw in the valley liner, consider the dynamics of flowing water. Sometimes water flows down one side of a valley with such force that momentum carries it up the other side of the valley, beyond the liner, and under the shingles. In this case a valley liner with a

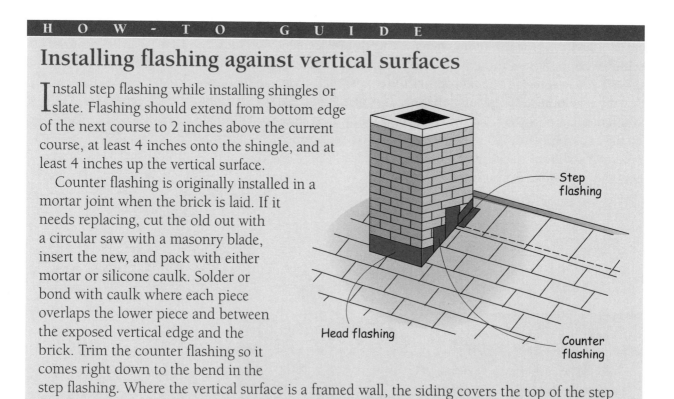

HOW-TO GUIDE

Installing flashing against vertical surfaces

Install step flashing while installing shingles or slate. Flashing should extend from bottom edge of the next course to 2 inches above the current course, at least 4 inches onto the shingle, and at least 4 inches up the vertical surface.

Counter flashing is originally installed in a mortar joint when the brick is laid. If it needs replacing, cut the old out with a circular saw with a masonry blade, insert the new, and pack with either mortar or silicone caulk. Solder or bond with caulk where each piece overlaps the lower piece and between the exposed vertical edge and the brick. Trim the counter flashing so it comes right down to the bend in the step flashing. Where the vertical surface is a framed wall, the siding covers the top of the step flashing and no counter flashing is needed.

Head flashing is fitted to brick the same way as counter flashing, caulked or mortared into a joint between bricks. Where the vertical surface is a framed wall, nail the top edge to the wall sheathing before installing siding. The bottom edge of head flashing should extend at least 4 inches onto the top course of shingles, covering the shingle nails.

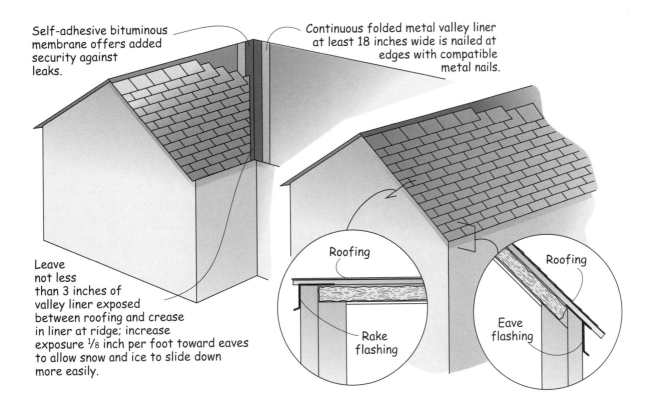

Self-adhesive bituminous membrane offers added security against leaks.

Continuous folded metal valley liner at least 18 inches wide is nailed at edges with compatible metal nails.

Leave not less than 3 inches of valley liner exposed between roofing and crease in liner at ridge; increase exposure ⅛ inch per foot toward eaves to allow snow and ice to slide down more easily.

Roofing

Rake flashing

Roofing

Eave flashing

8-6. Installing flashing in valleys and eaves.

ridge crimped into it at the fold in the center might solve the problem but can also hinder snow and ice from avalanching off. *(See Figure 8-7.)*

An alternative solution is to remove the shingles further back and lay down a wide strip of self-adhesive bituminous membrane often sold as "Ice and Water Shield." This material is gummy enough to seal around the nails holding shingles or flashing material over it. A 36-inch-wide strip will reach twice as far up the sides of the valley as the usual 18-inch-wide valley liner. With the membrane in place, install the usual valley liner and then the shingles or slates as you normally would. If that doesn't solve the problem then it's a fair bet that you're looking in the wrong place for the leak.

EXTRA HELP

To avoid corrosion problems, nail copper with copper nails; aluminum and stainless steel with stainless steel nails; and galvanized steel and lead with galvanized roofing nails.

For tight seals in problem areas, solder copper, galvanized steel, and lead. Seal aluminum with folded seams sealed with silicone caulk.

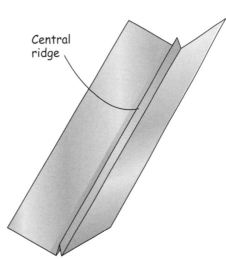

Central ridge

8-7. A central ridge crimped into a valley liner slows the momentum of water rushing down the adjoining roof and can help prevent water from rushing up the other side of the valley and under the shingles or slates on that side.

8-8. A sagging ridge indicates a problem with the barn's structure.

Installing new roofs

A new roof can mean many different things. It can mean stripping off the deteriorated shingles and putting new ones on; it can mean dismantling the entire roof structure down to the walls and building everything new from there on up; or it can mean something between those two extremes, including reinforcing or bracing the roof structure before putting on a new covering.

Sizing up a new roof job is like buying breakfast cereal; the choices are big, bigger, and biggest — there is no small. If your primary need is a new roof covering and you intend to strip off all the old covering and install a new one, you would be wise to reinforce or brace any areas that may be sagging while the old roofing is off. Removing the sag will be a lot easier while the weight of a roof covering is off and you have several good ladders on the site.

The most visible common structural problem in outbuilding roofs is a sagging ridge accompanied by walls that bow out. Since this is usually a problem with the main skeleton of the building and involves the walls as well as the roof, it was discussed in Chapter 5; the causes are shown in *Figure 8-9.*

Sometimes, however, the ridge remains straight but the rafters themselves are sagging. This may stem from inadequate sizing of the rafters when it was first built or from deterioration over the years. There are two effective approaches to solving the problem. You can replace or reinforce the rafters or you can give them additional support.

8-9. Bowed walls result from a break in structure across the barn.

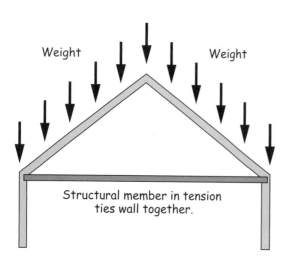
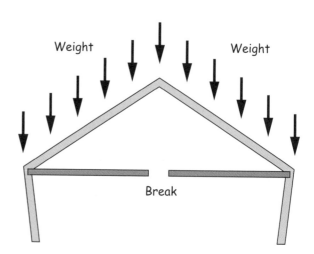

Structural member in tension ties wall together.

Break

Replacing rafters, particularly if many of them need replacement, will be easier if you first remove the roof deck. Then it's a straightforward matter of removing the old and fitting in the new.

Reinforcing a rafter in place can be a bit of a puzzler if you have to remove the existing sag. Often the sagging rafter has taken on a permanent bow, so you have to not only remove the weight causing the sag but also bend the rafter back to straight. You can do this by starting the reinforcing piece at one end, jacking the other end, and nailing the reinforcement to the existing rafter as you go. *Figure 8-10* shows the sequence.

If you want to add additional support to prevent further sagging but are willing to live with the existing sag, you can retrofit a purlin under the rafters. Just make sure the purlin has sufficient support. *Figure 8-11* shows how to do it.

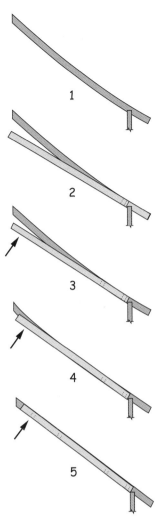

8-10. To straighten a bowed rafter (1) in place, first attach a partner at one end (2). Jack the reinforcing piece a the other end, nailing it every few feet to the rafter (3, 4, and 5).

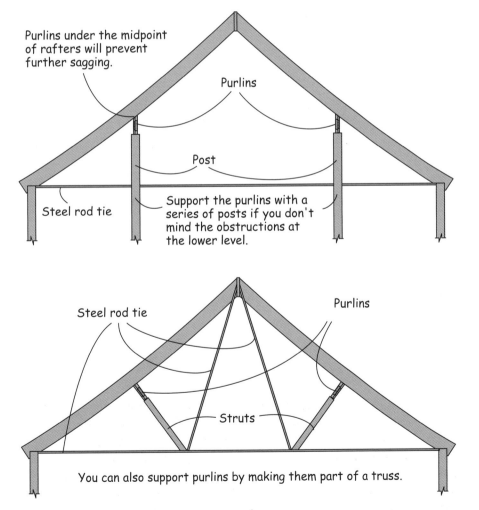

Purlins under the midpoint of rafters will prevent further sagging.

Purlins

Post

Steel rod tie

Support the purlins with a series of posts if you don't mind the obstructions at the lower level.

Steel rod tie

Purlins

Struts

You can also support purlins by making them part of a truss.

8-11. Two ways to support purlins.

Roofing weights

MATERIAL	WEIGHT (IN LBS PER SQ FT)
Asphalt shingles	2–3
Roll roofing, single coverage	0.9
Roll roofing, double coverage	1.3
Preformed metal	0.5
Standing-seam metal	0.75
Slate	7.5–40
Wood shingles	1.5
Wood shakes	3

In most cases you don't need to consult an engineer to determine if the structure is adequate. Your local building officials or an experienced builder can tell you how many pounds per square foot you need to allow for wind and snow loads in your area. The chart (*left*) will tell you how many pounds per square foot each layer now on the roof weighs. Your local lumberyard can tell you the load-carrying capacity of your rafters if you give them the dimensions. With those figures in hand you can calculate whether the roof will carry another layer. You'll notice from the chart that a metal covering only adds 0.5 to 0.75 pounds per square foot, while another layer of asphalt shingles will add 2 to 3 pounds. Using a lighter material could make the difference between stripping and not stripping.

You can even consider a new roof covering over slate. An experienced roofer that I know has been quite successful applying preformed metal over slate with power-driven gasketed screws.

Stripping a roof is not an advanced science, but you need to plan ahead to minimize the work involved. A few items to consider:

Where do you intend to dump the rubbish? There'll be a lot of it.

How will you get it there? As suggested below, a dump truck may be a much better choice than a dumpster.

How will you load it? If you can possibly arrange it, you'll save a lot of work if the material can go directly from the roof to the container, so that you can avoid having to pick it all up off the ground.

Do you have to protect foundation plantings or other equipment or structures under the eaves that cannot be removed?

How will you protect the building from the weather from the beginning of the stripping until the end of the reroofing?

Professional roofers like to back a dump truck right up to the building, under the eave. They can then push old material down the roof into the dump truck. After stripping the roof above the truck, they move the truck over for the next section. Where necessary, they may arrange a chute from just under the eave to the dump truck box. Dumpsters also work for this purpose. Old, stripped shingles are rough, messy, and full of nails, so you don't want to handle them any more than you have to.

With flat and light materials, such as old sheets of steel, make sure that you don't toss them off the roof only to have them sail back into the building or across the yard. To be safe, you may have to fold such materials into a more compact shape or drop them down a chute.

8-12. You can rent these waste-disposal containers and have them parked right under the eaves of the barn. A dump truck is an even more convenient option.

You can discard slate that has become crumbly the same as you would asphalt or wood shingles, but slate that is still sound is too valuable to throw away or risk breaking. If the slate is still good but is being replaced with a different material, offer it in a classified ad. In areas where slate is popular, roofers may come running to offer to take it off your hands (and roof). You might not come away with a roll of big bills, but at least you won't have to remove all that heavy slate and carry it down a ladder yourself.

The best tool for removing asphalt and wood shingles and crumbly slate is a special spade-like tool. The blade is slid up under the roofing as far as the fasteners, then the handle is pressed down to pull out the nails. You can use a spading fork in the same way if you can't get your hands on the specialized shovel. You'll also need a pry bar or framing hammer to remove the nails that remained in the deck.

Always remember: If you can pound a nail flush you don't have to remove it.

Metal roofing is a different story — it's impractical to pull out the fasteners by prying up the metal. Instead, you'll have to pull each fastener and then remove the sheet. To remove standing-seam metal roofing, you have to pry open the seams to get at the fasteners.

With the roof covering and underlay removed, examine the deck. Now is the time to replace any parts that have deteriorated. As you look it over, keep in mind that a roof deck is not a floor. With the exception of the roofers themselves, a roof deck supports very uniform loads. Those loads may be very heavy if you're in the snow belt, but they're spread evenly over the roof. A floor, on the other hand, may need to hold up a grand piano with all the weight concentrated at the bottom of three legs. The rule of thumb for roof decking is that it has to hold the roofing nails. If it's spongy and won't hold nails, replace it.

Roof sheathing

In the last half century roof sheathing has come to mean plywood. As with many other new products on the market, most people embraced

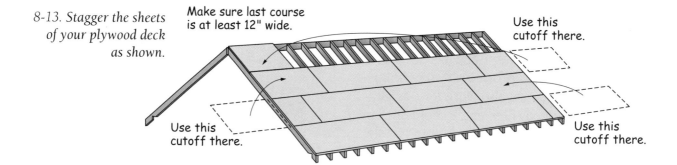

8-13. Stagger the sheets of your plywood deck as shown.

Make sure last course is at least 12" wide.

Use this cutoff there.

Use this cutoff there.

Use this cutoff there.

plywood for its advantages without thinking much about its disadvantages or the advantages of the products it replaced. Looking back we see a mixed record. Where plywood has replaced wood boards laid edge to edge under asphalt shingles or roll roofing, it has provided better bracing, better nailing, and a smoother surface. Where it has replaced strapping — thicker but narrower boards laid with spaces between them — under shingles, slate, or metal it has often resulted in a shorter useful life for the roof covering. The problem has been a lack of ventilation.

The need for ventilation of roof insulation and how to provide it was explained in Chapter 7 on page 164. Wood shingles or shakes, and to a lesser extent slate and metal roofing, also require ventilation immediately under them. If they are laid directly on a plywood deck then moisture condensing from vapor within the building or blown in through the joints in the roof covering has no way to quickly escape. In the case of wood shingles the result is cracking, mold, mildew, and rot. With slate the result is more rapid breakdown from freezing while damp. In the case of metal the result is more rapid deterioration of the sheathing. If the roof is sheathed with plywood, the only way to ventilate the underside of the roof covering is by adding strapping over the plywood.

If your building is large, and particularly if it's high as well, think also about how you're going to get material up there. Compared with strapping, plywood is awkward to carry up a ladder, and you can't climb it like a ladder after it's nailed down.

All things considered, I suggest that you use plywood with asphalt shingles or roll roofing, but I recommend strapping if you are roofing with wood shingles, slate, or metal roofing. If you strap, be sure to brace the roof well with diagonals nailed to the underside of the rafters.

Sheathing a roof will go fairly quickly and smoothly if you're organized and have an adequate crew. If the building is large, and especially if it's tall, like a big two-story barn, you should have two workers devoted to carrying the plywood or strapping up to the roof, two to position and nail the material to the rafters, and, once you get 12 to 16 feet from the eaves, two more to carry the sheathing from the eave to where it's needed. These last two can help out with either the carrying up or nailing down as needed. The job can be done with just two workers, but they seldom get into a really efficient rhythm because they're constantly switching between hauling material and laying it.

Nail a plywood deck with 8d (2½ inch) common nails 8 inches apart on each rafter. Space the sheets as recommended by the manufacturer, usually ⅛ inch apart. Begin alternate courses with half-sheets so that the joints going up the slope of the roof are staggered. (See Figure 8-13.)

If you're strapping the roof for metal roofing, space the strapping 16 inches on center starting at the eave. (See Figure 8-14.) Wood shingles and slates require closer strapping; use the same distance center to center as the exposure of the shingles or slates. Also, be sure to start strapping in the right place. The first strapping above the eave needs to be centered where the shingles or slates will be nailed. For wood shingles that distance will be equal to the exposure plus one inch. For slate, check where the slates are punched for nails and begin the strapping accordingly. Strapping should receive two 8d nails into each rafter.

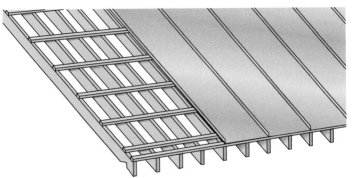

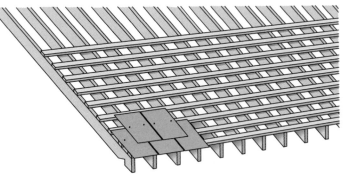

8-14. Strapping a roof can be done in two ways. For metal roofings (left), strap 16 inches on center with extra support at the eave and ridge. For wood shingles or slate (below), center-to-center spacing of strapping should be the same as the exposure of the shingles or slates. Make sure the first strap above the eave is centered under the fasteners.

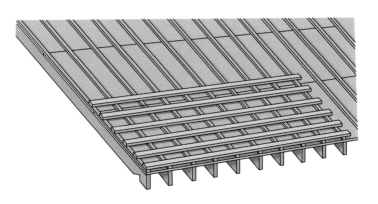

8-15. If a roof is already sheathed with plywood and you want cover it with wood shingles you will need to first strap it vertically to provide ventilation for the shingles, then horizontally to provide structure to nail the shingles to.

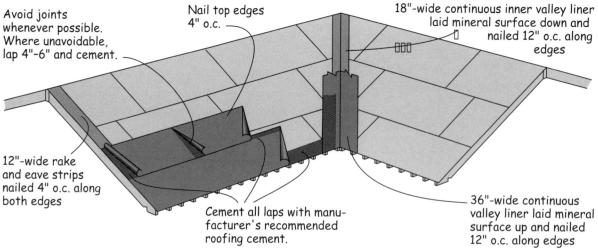

Avoid joints whenever possible. Where unavoidable, lap 4"-6" and cement.

Nail top edges 4" o.c.

18"-wide continuous inner valley liner laid mineral surface down and nailed 12" o.c. along edges

12"-wide rake and eave strips nailed 4" o.c. along both edges

Cement all laps with manufacturer's recommended roofing cement.

36"-wide continuous valley liner laid mineral surface up and nailed 12" o.c. along edges

8-16. Installing single-coverage roll roofing.

Roll roofing

Roll roofing is a heavy asphalt-saturated felt material with mineral grains embedded in the surface that will be exposed to the weather. Single-coverage roll roofing has mineral completely covering one side and is intended to be lapped 3 inches over the next course below. It can be nailed at both the top and bottom edges, with the top row of nails covered by the next course and the bottom row left exposed, or it can be nailed at the top edge and cemented to the lower course where it overlaps. (*See Figure 8-16.*) It is usually applied parallel to the eaves but can also be applied perpendicular.

Double-coverage roll roofing has mineral covering on only slightly less than half of the width and is a bit lighter. The next course overlaps the entire non-mineral portion. The top edge is nailed and the entire overlap is cemented. (*See Figure 8-17.*) Be sure to use roofing cement recommended by the manufacturer in the amount recommended. Too little cement invites wind damage but too much can cause blistering.

When a roof is covered with roll roofing it is usual to use the same material for valley liners and flashing. Begin by applying 12-inch-wide strips along the rakes and eaves. Then, if you have valleys, apply an 18-inch-wide strip down the valleys, nailing it along the edges just enough to keep it in place. Cover this with a 36-inch-wide strip, similarly nailed. (The nails holding the main covering strips are the primary fasteners for the flashing and valley liner.)

Lay the first course by aligning it with the edge of the eave flashing but nailing it only every couple of feet along the top edge. At valleys, trim the end 4 inches back from the center of the valley. Cut up from below to avoid cutting into the valley liner. When you're sure the course lies straight and flat, nail the top edge every 4 inches. Finally, while one worker holds the ends and edges back, the other coats the flashing with roofing cement. Press the ends and edges firmly into the cement and, if local conditions warrant, nail them while the cement is still soft.

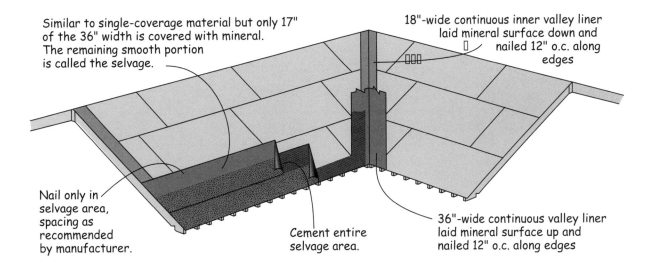

Similar to single-coverage material but only 17" of the 36" width is covered with mineral. The remaining smooth portion is called the selvage.

18"-wide continuous inner valley liner laid mineral surface down and nailed 12" o.c. along edges

Nail only in selvage area, spacing as recommended by manufacturer.

Cement entire selvage area.

36"-wide continuous valley liner laid mineral surface up and nailed 12" o.c. along edges

8-17. Installing double-coverage roll roofing.

Apply succeeding courses in the same manner, overlapping the previous course of roofing by at least 2 inches and preferably 3 inches for single-coverage roofing, or down to the mineral-coated portion for double-coverage roofing.

Asphalt shingles

Asphalt shingles are made of the same materials as roll roofing — an asphalt-saturated fiber with mineral granules embedded in the exposed surface. They are sold as shingle strips, where each strip has cutouts along the mineral-surfaced edge to form tabs that were originally intended to mimic the appearance of wooden shingles. Today, asphalt shingles have so overtaken the roofing market that most people no longer think of shingles as being made of wood. If you mention shingles to them they think only of asphalt.

Asphalt shingles are nailed over a 15-pound underlayment into the roof deck. Valleys are often lined with matching roll roofing, first an 18-inch-wide strip laid mineral surface down and then a 36-inch-wide strip laid mineral surface up. *(See Figure 8-18 on next page.)*

Begin installation with a preformed metal drip edge at the eaves. Then lay the 15-pound underlayment in strips parallel to the eaves. The underlayment will function just as well if you lay it perpendicular to the eaves, but the added thickness where it overlaps will be apparent through the asphalt shingles after a few warm days with hot sun. That irregularity is camouflaged when it runs parallel to the butt edges of the shingles. If there is any chance of wind before the underlayment is covered by shingles, fasten it down with cap nails, which are small ring-shank nails with plastic discs under the head. The underlayment is much less likely to tear free from a cap nail than from an ordinary roofing nail. Finally, install preformed metal drip edge along the rakes.

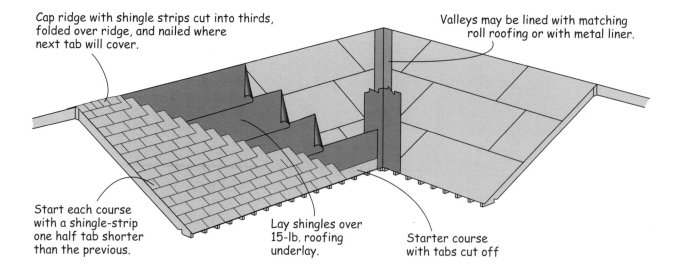

Cap ridge with shingle strips cut into thirds, folded over ridge, and nailed where next tab will cover.

Valleys may be lined with matching roll roofing or with metal liner.

Start each course with a shingle-strip one half tab shorter than the previous.

Lay shingles over 15-lb. roofing underlay.

Starter course with tabs cut off

8-18. Installing asphalt shingle roofing.

In colder climates where there is any potential for ice dams forming on the edge of the roof, install a 3-foot-wide strip of self-adhesive bituminous membrane along the eaves. If the roof overhang at the eaves is more than a foot or so, install two such strips overlapping 2 or 3 inches. This ice and water shield may be unnecessary if the roof is adequately ventilated, but the cost for extra protection is small compared to the possible expense of repairing water damage later — and, yes, I do wear both belt and suspenders.

If there are valleys in the roof they must be lined before you can start actual shingling. The roll roofing liner mentioned above is the most common liner, but a metal liner sheds leaves, ice, and snow more readily. If you're reroofing and the existing metal valley liners are in good condition, then you don't have to replace them; run the underlayment and the new shingles right over them. You can also put a new valley liner over an old one.

Shingling begins with a starter course of shingle strips with the tabs cut off. The first one should also have 3 inches or so cut off one end so that the joint with the next one won't line up with a cutout in the next course. Nail the starter course about 3 inches above the eave, spacing the nails about one every foot. You may want to nail the starter course as you install the first full course in order to make sure the starter course nails don't wind up right under a cutout in the first full course.

Begin the first full course with a full-length shingle strip aligned with the starter course at the eave. Asphalt shingles come in a variety of configurations, so check the manufacturer's recommendations for nail placement. Typical instructions call for a nail about ½ to ¾ inch above each cutout and one about 1 inch in from each end. When you're still a good distance away from the other end of the course, check how the shingles are going to come out. Adjust the spacing between strips so you come out with either a full-tab or a half-tab at the end.

Begin the second course with a shingle strip from which you've cut off a half-tab so the center of the tabs align with the cutouts in the shingle strip below. As you work up the roof remove an additional half-tab to start each course so the joints will be well staggered.

When both sides of the roof are shingled you have to cap the ridge. If the space under the roof is ventilated at the gable ends you can do this with tabs cut from shingle strips and folded over the ridge. Start at the downwind end and work into the prevailing winds, nailing the tabs where the nails will just be covered by the next piece. If your ventilation arrangements call for a vented ridge, use a preformed ridge vent/cap and install it according to the maker's recommendations. It is best if you are able to use gasketed screws for installation, because strong winds tend to pull out nails holding metal ridge caps.

8-19. Barn roofs are frequently an opportunity for folk art. Owners often apply shingles in a mosaic to create a picture or spell out a message.

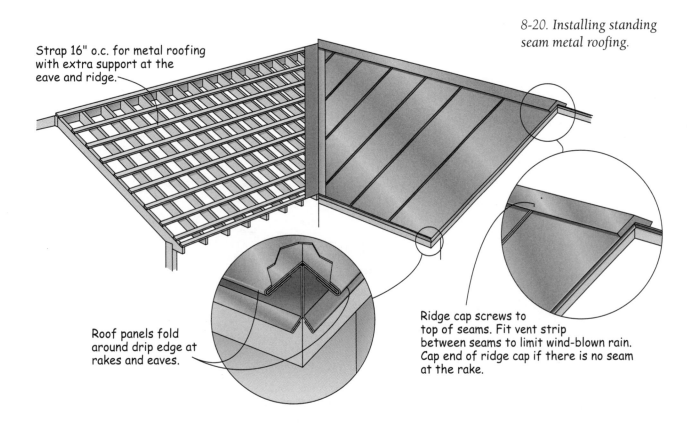

8-20. Installing standing seam metal roofing.

Strap 16" o.c. for metal roofing with extra support at the eave and ridge.

Roof panels fold around drip edge at rakes and eaves.

Ridge cap screws to top of seams. Fit vent strip between seams to limit wind-blown rain. Cap end of ridge cap if there is no seam at the rake.

Metal roofing

Metal roofing has come a long way in the last few decades. It no longer should conjure up the old images of partially rusted, galvanized-gray, waveform corrugations on cheap industrial buildings. Baked-on and porcelain enamel finishes over galvanized steel or aluminum have added decades of extended life and a rainbow of colors to this lightweight, economical, and easily installed roofing. And, unlike asphalt roofing, when metal roofing finally does have to go it can be recycled instead of tossed in a landfill.

The premium metal roofing, standing seam, is now a much more competitive option, thanks to portable forming machines. *(See Figure 8-20 on page 201.)* While it's still impractical for the inexperienced, it no longer requires hours of skilled hand-forming of seams. Instead, the metal is delivered on a small truck to the site in a coil, and it is fed directly into the forming machine that came trailed behind the truck. In an hour, all of the panels for a large roof are formed and cut to precise length.

Metal roofs are especially advantageous in areas that have high snowfall. Snow slides off of them so readily that snow loads are vastly reduced.

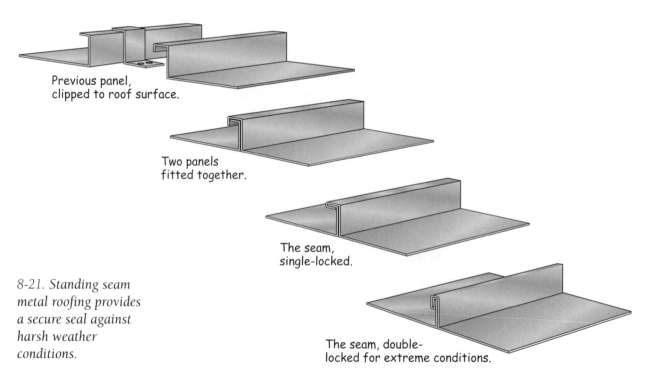

Previous panel, clipped to roof surface.

Two panels fitted together.

The seam, single-locked.

The seam, double-locked for extreme conditions.

8-21. Standing seam metal roofing provides a secure seal against harsh weather conditions.

Two precautions need to be mentioned. First, the economy of installing metal roofing can disappear quickly if the roof is complicated. Metal roofing goes up much faster than other forms of roofing on straight runs with no penetrations, but complications like valleys, hips, dormers, and vents require much more time to work around. Second, in some situations condensation on the underside of the roofing can be a problem. If your building is likely to have high humidity inside on cold nights, you might want to consider sheathing the roof with inexpensive fiberboard and a layer of 30-pound underlayment. This combination will effectively prevent dripping condensation.

Preformed metal roofing differs from standing seam in how it is fastened to the roof. Although there are now snap-together forms of preformed material, the usual forms are nailed or screwed with gasketed fasteners from the top side. The fasteners holding a standing seam roof are all under the metal — weather can't penetrate the surface. This is achieved by means of a metal clip that is nailed to the deck or strapping at the working edge. The clips are then folded around the standing edge and covered by the next panel. The seam is completed by folding the edge of the new panel around the standing part of the previous panel, locking the clip into the seam.

The do-it-yourself outbuilding restorer wanting a metal roof should choose preformed panels. (*See Figure 8-22.*) These can be bought from inventory in a range of lengths or can be ordered cut to your requirements. Metal roofing is normally continuous from eave to ridge unless the

8-22. Installing pre-formed metal roofing.

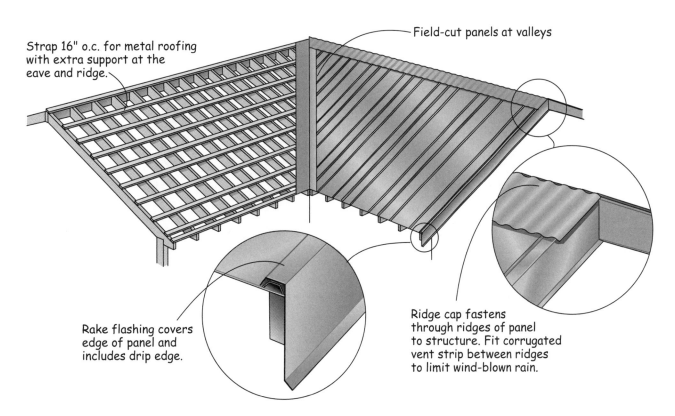

Strap 16" o.c. for metal roofing with extra support at the eave and ridge.

Field-cut panels at valleys

Rake flashing covers edge of panel and includes drip edge.

Ridge cap fastens through ridges of panel to structure. Fit corrugated vent strip between ridges to limit wind-blown rain.

distance is so great that handling the panels or expansion and contraction would be a problem. Keep in mind that metal expands and contracts with changes in temperature. A roof can easily reach 150 degrees Fahrenheit under a summer sun and can easily drop to 30 below on a clear January night in northern areas. That's a range of 180 degrees. With only a 100 degree temperature range, a 30-foot panel of aluminum roofing will expand and contract nearly ½ inch over that length. Galvanized steel roofing will expand and contract half as much. You should count on half of the expansion pushing up toward the ridge as the other half pushes down toward the eave. If you think you might be looking at a potential problem, talk to the roof's manufacturer.

Installation is straightforward. Preformed metal is installed with a 1-inch overhang at the eave, becoming its own drip edge. At the gable ends a matching rake-edge flashing covers the edge of the metal roof, extends down over the trim, and finishes with a drip edge. At the ridge, a closure flashing wraps around the ends of the panels and then a ridge cap covers it all. The usual fastener for all of it is a screw with a neoprene washer or gasket, power driven. It's all very neat, and quick.

Doors and windows

We ask a lot of doors and windows, so it should come as no surprise that good ones are highly engineered and expensive. And considering how highly engineered they are, and how much use and abuse they take, it should come as no surprise that they occasionally fail. Even the relatively simple ones found in most barns and outbuildings fail because of the harsher environment that they're subjected to.

When restoring an outbuilding you therefore need to first consider the severity of the building's environment, and then decide which are the essential features of the doors and windows you're considering repairing or replacing.

Fixing door and window problems

Doors and windows in unheated outbuildings have some problems that are different from those in heated buildings. Sticking in the summer and then contracting and leaking air in the winter are less common problems because the winter air isn't as dry as it is in a heated building. On the other hand, it is more common for window and door openings in barns and outbuildings to become out of square. This is usually a result of shifts in the building's structure, which may occur because an outbuilding's foundation is often not very deep and there is usually no heat to keep frost out from under it. Rot is also more common, probably because outbuildings are often not that well maintained or may not be painted as regularly as a house.

Doors in crooked buildings

A door opening that is out of plumb (not perfectly vertical) or out of square is almost always the result of structural problems — either the foundation is sinking or the building is tilting. Fixing the out of square problem without addressing the cause will be a short-lived repair. However, if you've taken steps to prevent further movement of the building and want to fix the opening without straightening the building, there are steps you can take.

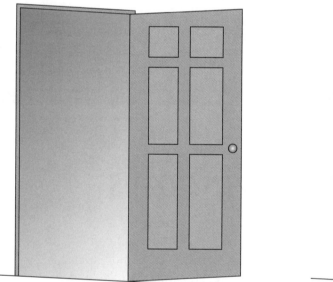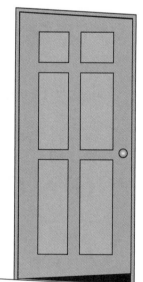

9-1. A door must hinge to a vertical jamb to prevent it from being opened or closed by the force of gravity. You can trim the door at the top and bottom to fit an out of square opening, provided you do not cut into the structure of the door and you find the trimmed door's appearance acceptable.

If the door swings downhill (left), it may look a bit odd, but it will still fully close the opening. If you trim the door to clear the floor when swinging uphill, however, it will leave a gap when closed (right). Closing the gap with a tapered threshold would be hazardous.

Keep in mind that a door must be plumb, or at least the hinge edge of the door must be plumb. If it's not, gravity will swing the door for you. To fix a door that is out of plumb, remove the door and jambs and reinstall the jambs plumb. If the wall tilts in or out you will probably have to use new, wider jambs. If it tilts left or right you will have to trim the door to fit the new opening, or even replace it with a narrower door. Tilting walls are often accompanied by sloping floors. Doors on plumb jambs present a knotty problem when they swing over such floors. If the door is in an exterior wall, the simplest and usually most satisfactory solution is to make sure the door swings outward, as there will almost always be a step down to provide clearance for the bottom of the door. If the door absolutely must swing in, the best solution is to hang the door on the downhill jamb. It will look odd but it will behave itself. The toughest situation is where the door must swing in and must hang on the uphill jamb. There is no way to avoid a gap at the bottom of the closed door. If the door is only intended to keep out cows, then the gap won't be too much of a problem. If you are trying to keep out blowing snow, then you're out of luck. A tapered threshold might fill the gap but you would constantly trip over it.

9-2. *When trimming a door to fit a doorsill, be sure it will also clear any sloping floor areas beyond.*

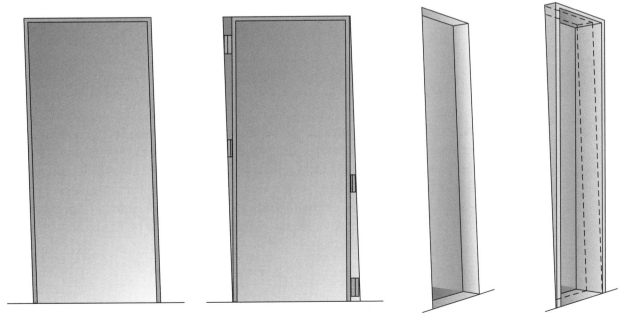

9-3. *If the opening tilts left or right, reinstall the jambs, shimming them plumb, and then install a narrower door.*

9-4. *If the wall tilts in or out, install wider jambs and threshold so you can hang the door plumb.*

Having fixed the opening, you will most likely have to trim the door to fit the new shape and size. Here you will probably have an easier time of it than your neighbor who's restoring an old house. His doors are likely to be panel doors that will be seriously weakened if too much is trimmed from either the stiles or top or bottom rails. Your outbuilding doors, on the other hand, are probably built of boards held together with cleats and braces; doors of this type are much less likely to be seriously weakened by aggressive trimming and can even be disassembled and rebuilt if necessary.

Windows in crooked buildings

Windows present a different set of problems when the walls that they're mounted in are out of kilter. On the negative side, the window itself is not that likely to have survived in usable condition if the wall has tilted so the opening is out of square. But on the positive side, windows usually function normally even if the opening is out of plumb. In New England, or at least in Vermont, it is quite common to see double-hung windows installed at a marked angle to fit into the wall area between two roofs, as shown in *Figure 9-5*. The window does need to be flat, however. It's not particularly common for a wall to be very far out of flat, but it can happen. If it does, and the window has survived but doesn't work properly, remounting and shimming as necessary will fix the problem. If you had the same problem with a door, simply adjusting the stops would do.

When a wall tilts enough that the opening is out of square the only remedy is to remove the window unit and square up the opening. (Unlike with a door, you can't trim a window unit to fit a smaller opening.) If the movement has been very small and the rough opening was made generously oversized, you may get by with adjusting the shims between the rough opening and the window. More likely you'll have to plane or chop away parts of the header and sill and perhaps the studs. If the opening is so far out of square that planes, chisels, and draw knives won't do the job, then it's highly unlikely that you will be able to put the same window unit back anyway — in that case, you'll be rebuilding the opening for the replacement unit.

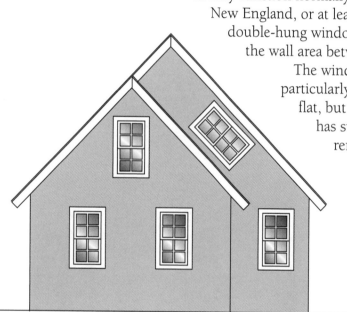

9-5. Practical New Englanders have known for years that windows don't need to be plumb to function well. The Vermont-style window at top right makes a striking feature.

Rot

The only thing good about rot in windows is that when you discover it you can angrily call its name without having to use a four-letter word. Unfortunately, rot happens. Where it happens is usually even more unfortunate — in the joints between two pieces of wood. That means that both

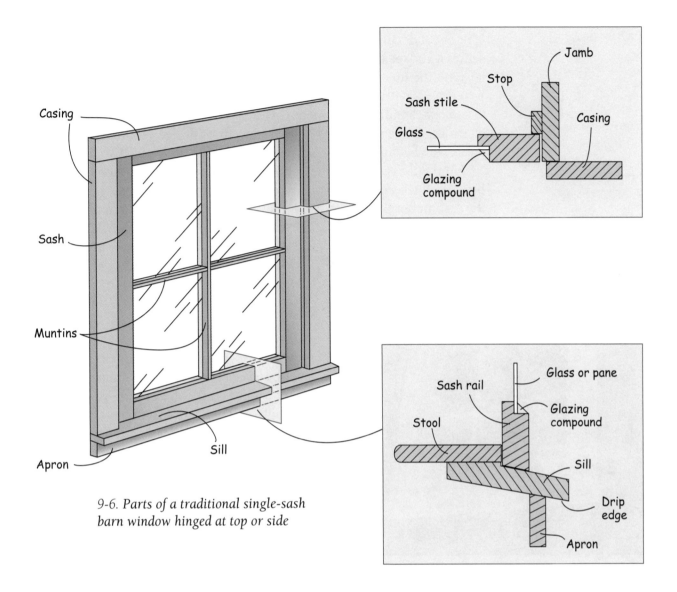

Casing

Sash

Muntins

Apron

Sill

Jamb

Stop

Sash stile

Glass

Casing

Glazing
compound

Glass or pane

Sash rail

Glazing
compound

Stool

Sill

Drip
edge

Apron

*9-6. Parts of a traditional single-sash
barn window hinged at top or side*

pieces are affected in the most critical location, right where you need
strength and fasteners. Before you jump into fixing the window, you
should do two things.

First, examine the situation and try to figure out just why it hap-
pened. Wood rots when there's excessive moisture in it. Where did the
moisture come from and how did it get into the wood? The source may
be condensation, though that's more likely in a heated building than an
unheated outbuilding. It may be from a leak in the roof, in which case
there should be other evidence of water damage nearby. Most likely, of
course, it's rain and snow. Then how did it get into the wood? Did it find
its way into the cracks and spaces around and under the unit or in
through joints between the members of the unit itself? The most likely
answer is that rainwater has seeped in through joints in the unit. You may
now beat up on yourself for having gone fishing that weekend when you
should have been painting the barn windows. Just remember that win-
dowsills rot — always.

Second, consider whether attempting to repair it really makes good sense. If the rot is entirely in the window frame members, the jambs and sill, and the unit is a very simple design, you may be able to replace the rotten parts without taking up your entire two-week vacation to do it. Take very careful measurements before you begin to disassemble it, then check them after disassembly. Reproduce the affected members (or the entire frame while you're at it) and reassemble it.

If the rot is in the sash itself (a likely spot would be in the joint between the bottom rail and a stile), then the only good reason to repair the sash rather than replace it is that the building is of substantial historic interest and the windows are original. If this is your situation, an amateur repair job won't cut it — ask your local historical society to refer you to a professional. The pro should be familiar with the National Park Service's Preservation Brief 9, "The Repair of Historic Wooden Windows." (If you are interested in learning more about proper historic window preservation yourself, a version of this document is not hard to find on the Web.) In most cases, though, you're dealing with a pretty charming outbuilding but not a national treasure, and you just want to preserve its original appearance. Ask at your local lumberyard for the sizes of barn or utility sash that they can order for you. If nothing matches what you need, look for a shop that reproduces window sash; they're surprisingly not that hard to find. If the shop is properly equipped, it will be able to reproduce typical outbuilding sash for a reasonable price. If you're still coming up blank, you may have to decide to replace the entire unit. Just don't take on repairing or replacing the sash yourself unless you know for sure that you have the woodworking experience and tools to do the job.

9-7. Barn windows are often of simple single-sash construction, as they are opened and closed less frequently than house windows.

Installing manufactured doors and windows

Modern prehung doors and window units are not foolproof, but they are a lot quicker and easier to install than doors and windows of a few decades ago. For the most part, there is no cutting or planing, no mortising for hardware, and often no tedious painting — just set them level and plumb in their openings and nail them to the wall.

Installing windows

Window units with attachment flanges around the outside edges are now more common than traditional units with wood casings already attached, but installation is mostly the same. Nevertheless, read the manufacturer's instructions carefully before beginning. In broad outline, the procedure will go something like this: Check that the rough opening is the correct size for the unit and that it is quite close to plumb and level, then place the window unit in the opening from the outside. While a helper holds the unit in the opening, shim the two lower corners from the inside so it is perfectly level and centered from left to right, then check that the unit is square by measuring the diagonals. If necessary, shift the bottom from side to side to square the unit. Nail the flanges to the wall at the bottom two corners, then the top two corners. Check that the four sides are not bowed in or out, shim them as necessary, and finish nailing the flanges as specified. That's nearly about all there is to it.

9-8. This barn was completely refitted with manufactured doors and windows to create an office building.

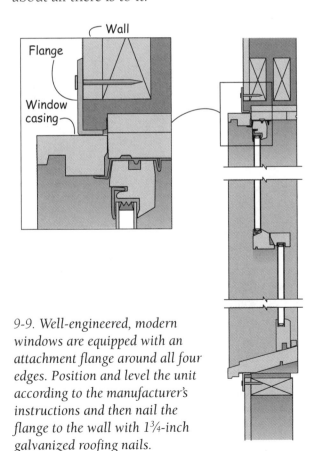

9-9. Well-engineered, modern windows are equipped with an attachment flange around all four edges. Position and level the unit according to the manufacturer's instructions and then nail the flange to the wall with 1¾-inch galvanized roofing nails.

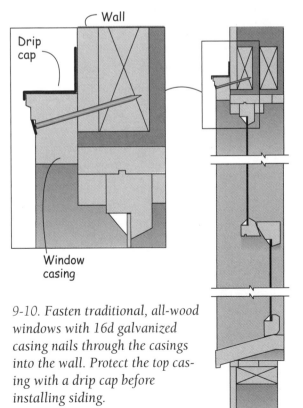

9-10. Fasten traditional, all-wood windows with 16d galvanized casing nails through the casings into the wall. Protect the top casing with a drip cap before installing siding.

Making your own windows

Modern manufactured window units are expensive. They are well worth the price for residential or office spaces, but they may be more than you need for spaces that do not require heating or cooling. Building your own windows using readily available lumber and parts is an economical alternative.

Outbuilding windows

The following design puts economical outbuilding windows within the grasp of just about any do-it-yourself restorer. The windows admit light and allow you to see out, they open to provide ventilation, and they easily remove without tools for cleaning or repainting. While they may be a bit drafty in a big blow if you don't weatherstrip them, they provide all the protection needed for most situations. It is hard to imagine how a window could be more economical unless you have a source for suitable used windows at an especially good price.

This design is particularly economical because it fits into walls with studs 2 feet on center without any additional framing. If your situation requires a window with other dimensions, simply frame an opening ½ inch wider and higher than the sash you intend to use. (Ask for "utility" or "barn" sash at your lumberyard.) You could also build a conventional frame for the window for a more finished look.

For each window of the design as shown you will need:

- One 6-light barn sash, 22 inches wide by 41¼ inches high
- One 2×6, 22½ inches long
- One 2×8, 22½ inches long
- 12 feet of 1×2
- 1 gate hook
- A small handful of 16d common nails and a handful of 4d box nails

If you really want to splurge you can lay out a buck and a half for a spring latch at the top of the sash instead of a quarter for a gate hook at the bottom.

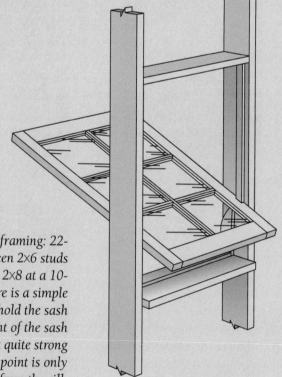

The sash is hung directly in the framing: 22-inch-wide wide barn sash between 2×6 studs 24 inches on center. The sill is a 2×8 at a 10-degree angle. The only hardware is a simple gate hook at the bottom to hold the sash closed. When open, the weight of the sash holds it in position in all but quite strong winds. Note that the pivot point is only about a third of the way up from the sill.

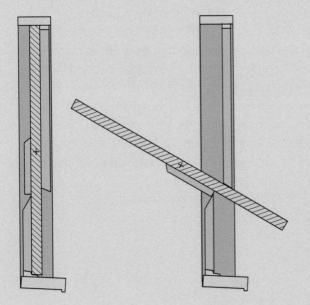

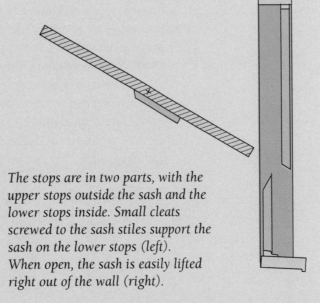

The stops are in two parts, with the upper stops outside the sash and the lower stops inside. Small cleats screwed to the sash stiles support the sash on the lower stops (left). When open, the sash is easily lifted right out of the wall (right).

The drawings show how everything goes together. Begin by checking that the studs are on 2-foot centers and provide at least 22¼ inches of space between. If your studs are rough-sawn and a full 2 inches thick, leaving you with only 22 inches between, you'll have to plane the edges of the sash to fit. Install the sill first, at about a 10-degree pitch sloping out, nailing through the studs into the sill with 16d nails.

Then cut and install the lower stops. The ones for the sides should be 14 inches long (roughly one-third of the height) with a 60-degree bevel at the top; cut the one at the bottom to fit between the ones on the sides.

Now hold the sash in place against the stops, resting on the sill, and mark the studs ½ inch above the top edge of the sash. Set the sash aside, square across the studs at the marks, and install the header 2×6 just above the marks.

Put the sash back in place against the lower stops and shim it up ¼ inch from the sill. While a helper holds it against the stops from the outside, mark the sash stiles at the tops of the stops from the inside. Cut two 6-inch pieces of 1×2 and nail them just above the

marks on the sash stiles. Support the sash from below and nail carefully.

Put the sash back once again and pivot it to the closed position. Mark the studs and header at the outside edge of the sash. Now pivot the sash 60 degrees and mark where the bottom of the outside stops must come.

Cut and nail in place the top outside stop. Now cut the upper side stops to fit exactly from the top stop to the marks indicating the bottoms of the side stops. The cut at the bottom should be at a 30-degree angle. With the sash in place and held tightly closed, hold the upper side stops against the sash and nail them to the studs.

Finally, install the gate hook at the bottom, or, if you've splurged in the hardware department, mortise and install the spring catch at the top.

If you want a bit more sophistication than this design and have the woodworking skills and tools, you can make your own hinged casement window units. Buy utility sash and build the window frame as shown in *Figure 9-6* on page 209. Hinge the sash to a jamb and provide a turn-latch to keep it closed.

When you are working on an old outbuilding, there are a few things to keep in mind. First, don't twist a window. If the wall isn't flat, shim out the flange or casing as necessary to keep it flat, fasten it through the flanges or casings and shims, then flash carefully to keep moisture out. Self-adhesive bituminous membrane tape 3 to 6 inches wide, if applied in warm weather, can be molded to pretty irregular surfaces and will do a good job of keeping out the elements and preventing rot.

Second, pay attention when checking the rough opening. If it's smaller than the recommended opening either the unit won't fit or you won't have any room to adjust it within the opening. But if it's bigger then there's a risk that the nails will be dangerously close to the edges of the framing, or miss the framing entirely. An out of square opening will usually result in the disadvantages of both too large and too small windows — some places will be too tight to fit and other places will have gaps that are too large.

9-11. To install a manufactured door or window, simply set it in the opening, then shim it in place with wooden wedges so the casing is square. Drive nails or screws through the casing and the wedges into the frame.

Third, think through the exterior wall covering before you install the windows. These units are designed to be installed over the sheathing but before the siding goes up and with siding that is no more than about ¾ inch thick. These conditions are not necessarily possible when restoring an outbuilding. You may need to incorporate some "unique architectural features" to adapt these units to the job at hand. Just be sure you are as creative at keeping out the rain as you are at adapting the building to accept these new windows.

Hanging doors

If you need a high-quality, weather-tight, exterior door, buy a pre-hung unit. That advice is not based on a low estimation of your wood-working skills but rather on a high opinion of the sophisticated weatherstripping that is found on many such doors and that would be difficult to install on-site. If economy is a prime consideration, look at prehung steel doors. Many of these use magnetic weatherstripping just like your refrigerator. Buy the door "prepped," meaning pre-bored (mortised) for the lockset.

Prehung doors do not have mounting flanges like window units but many do have brick-mold casings already installed on the exterior. Don't be misled into thinking that you can simply nail the casings to the wall like a window, however. Doors are heavy and subject to much more stress and abuse than windows — the jamb that carries the hinges in particular must be securely fastened to the building's framework.

Before you begin the actual installation, check the rough opening. The bottom of the opening must be level, flat, reasonably smooth, and at the correct height so that the door bottom will clear the inside finish floor. If necessary, shim the bottom of the opening with plywood of appropriate thickness set in a bed of construction adhesive or suitable caulk and then well nailed. The sides of the opening also need to be sufficiently flat and smooth so that shims between the opening and the jambs will be well supported. You don't want to distort the jambs when nailing them to the building's structure.

To install the door unit, place it in the opening and center it. Adjust the position so that the hinge-side jamb is perfectly plumb both from left to right and from inside to outside. When you're satisfied that the unit will fit properly, remove it and apply a heavy bead of caulk on the bottom of the opening along the outside edge, then place the door unit back in the opening. Fit shims between the opening and the hinge-side jamb opposite all three of the hinges and at the bottom. Make sure the jamb is plumb both ways and fasten it with 16d galvanized casing nails or 3½-inch galvanized drywall screws, two at each shim position.

Now turn your attention to the lock-side jamb and put your level aside — you won't need to use it anymore. Shim the lock-side jamb so the clearance between the closed door and the jamb is uniform from top to bottom and the weatherstrip between the door and the stop is uniformly compressed along its entire length. You should have shims at the bottom, near the top, and opposite the latch hardware. When you've got it right, fasten it as you did the opposite side. For a really secure installation, go back to the hinged side, remove one or two of the screws holding each of the hinges to the jamb, and replace them with screws of the same gauge but long enough to reach all the way into the framing. If the lockset does not already have provisions for fasteners that go into the framing, replace the strikeplate screws with longer ones as well.

Board and batten doors

Outbuilding doors that will keep out animals and the bulk of nasty weather are not difficult to make and will last a long time if you keep a couple of things in mind. First, make sure your door has adequate diagonal bracing so it won't sag in a couple of years. Second, hang it with the battens and bracing on the inside so rain and snow won't collect on the upper edges and invite rot.

Traditionally, this type of door was assembled with clinched nails. Good secure fastening is necessary, and a straight nail that is short enough to not protrude when fastening two boards together is too short to do the job. A properly clinched nail, however, is not one that is simply hammered over on the back side; it is one that curls back into the wood after exiting the back side. To do this, place an axe head under the assembly where the nail will come out, then nail at a slight angle. As the nail comes out it hits the axe head and is deflected. As you pound it further in it continues to deflect, curling back into the wood. You won't have to worry again about tearing your jacket every time you brush against a hammered-over nail. These days it's more common to see batten doors assembled with power-driven galvanized screws.

9-12. The Dutch door, shown here with diagonal bracing, is a classic barn feature.

There are two ways to approach bracing the door. You can install the boards on the diagonal so they provide the bracing or you can use a separate diagonal batten. (*See Figure 9-13.*) Ignoring appearance, which may make the choice for you, both approaches have advantages. Diagonally boarded doors use boards of various lengths, so they may make more efficient use of the materials you have available. Straight, vertically boarded doors, on the other hand, are usually quicker and easier to make. In either case tongue-and-groove boards will make a more weather-tight door; just make sure you install them tongue-side up if you're using them diagonally.

Making and installing barn doors

Barn doors are big, big enough to admit a team of horses pulling a hay wagon or, more likely today, a farm tractor with a big implement in tow. They're so big that hanging them from hinges puts enormous strain on the door and the wall they're hinged to; so big that just a moderate breeze makes them a menace. Years ago barn doors were hinged because there was no practical alternative. For the most part, they have all succumbed to the destructive force of their own weight and fickle winds. The practical alternative ever since the industrial revolution has been a barn door hung from roller hangers riding a track. Consider no other type of wide barn door unless it is an overhead garage door, which is not something you can make yourself.

The barn door is best built flat on the barn floor, then raised up like a section of wall and hung from the track. *(See Figure 9-14 on page 219.)* You can frame doors up to 50 square feet or so with 2x framing on the flat, making a finished door about 2 to 2½ inches thick after covering the framework with siding.

EXTRA HELP

If you happen to live in an area where the catalpa tree grows, ask a local sawyer where you can buy catalpa boards. It is an extremely stable wood, exceptionally rot resistant, and soft enough to be easy to work with — characteristics that make it ideal for making your own doors and windows.

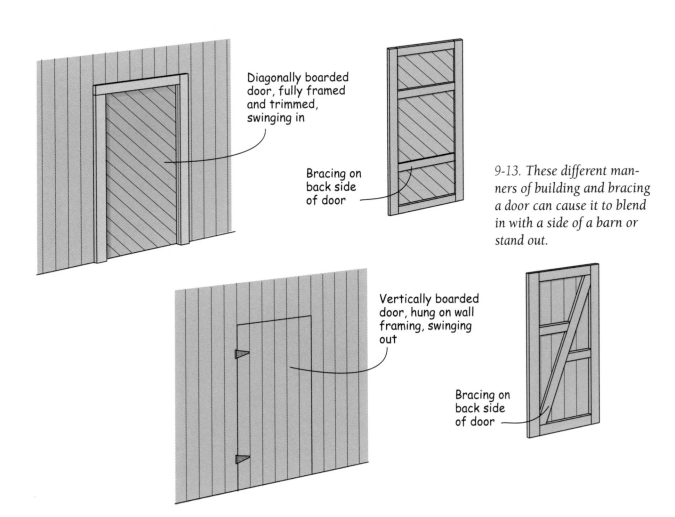

Diagonally boarded door, fully framed and trimmed, swinging in

Bracing on back side of door

9-13. These different manners of building and bracing a door can cause it to blend in with a side of a barn or stand out.

Vertically boarded door, hung on wall framing, swinging out

Bracing on back side of door

Pick out your straightest
and flattest boards for the
barn door.

You are best off framing a larger door with the framing lumber on edge, like a stud wall, but running the framing horizontally. Fasten the framing together well, using reinforcement plates at the joints if you're framing the door on the flat.

The key to a smooth and successful barn door project is planning. Buy the hardware, all of it, before you start driving nails. Make sure the roller hangers that you have were made for use with the rail that you have. Lay out the rail and hangers and study any accompanying instructions so you understand what clearances are needed. Make sure you understand where you will attach the bottom guide or stay roller and how much the door must overlap the adjoining wall for this bottom hardware to function properly. And make sure you will be able to securely lock the doors.

Don't omit a stay roller or bottom guide under the misconception that the weight of the door will keep it from blowing out from the building in a storm. A good wind can blow a 400-pound door nearly horizontal, and even if the door doesn't come loose from the rail it will smash back into the building when the gust stops.

HOW-TO GUIDE

Making a door to match an old barn

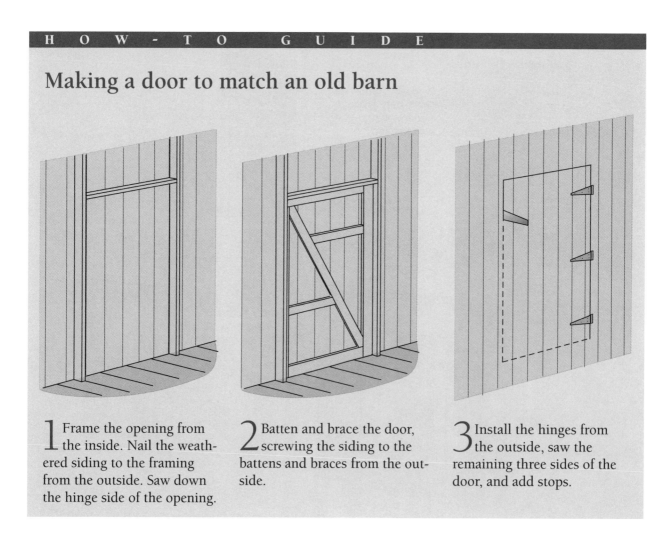

1 Frame the opening from the inside. Nail the weathered siding to the framing from the outside. Saw down the hinge side of the opening.

2 Batten and brace the door, screwing the siding to the battens and braces from the outside.

3 Install the hinges from the outside, saw the remaining three sides of the door, and add stops.

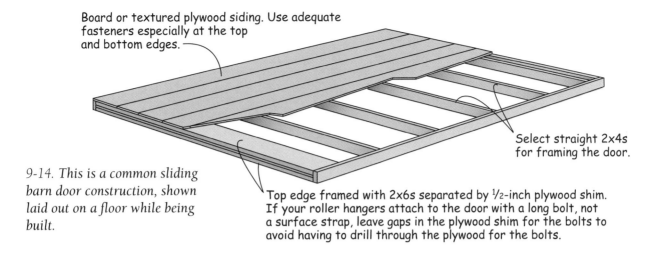

Board or textured plywood siding. Use adequate fasteners especially at the top and bottom edges.

Select straight 2x4s for framing the door.

9-14. This is a common sliding barn door construction, shown laid out on a floor while being built.

Top edge framed with 2x6s separated by ½-inch plywood shim. If your roller hangers attach to the door with a long bolt, not a surface strap, leave gaps in the plywood shim for the bolts to avoid having to drill through the plywood for the bolts.

9-15. This is a typical 24 × 24-foot outbuilding with "hayloft" access from outside and a 12-feet-wide by 8-feet-high ground-level doorway closed by two 6-feet-wide doors rolling on a single 24-feet-long track. Hardware for the loft consists of three T-hinges and a hook; for the ground-level doors, two 12-feet lengths of track, track hangers, two pairs of roller door hangers, and two stay-rollers to keep the doors from blowing out from the building.

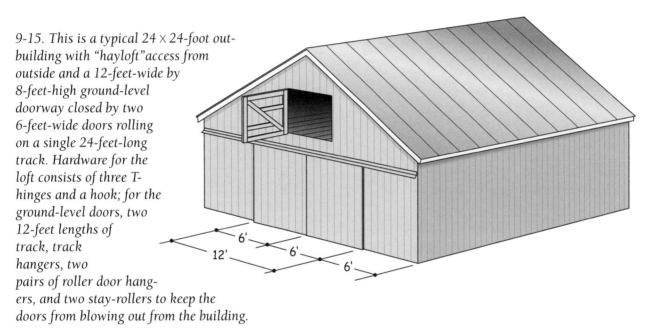

6' 6'

12'

6'

EXTRA HELP

To preserve the historic appearance of an old barn, look around for old-style track and rollers. A metal working shop can straighten out old track and even repair roller wheels. Grease up the wheels and track, and the restored parts will work like new.

Using old hardware

Some parts of a barn's doors will need to be replaced because of rot, rust, or plain ineffectiveness. But whenever you can retain a charming, authentic hardware detail, be sure to do so.

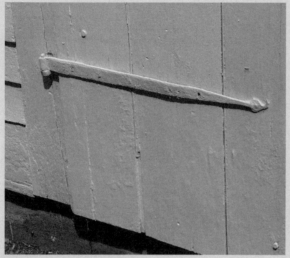

A traditional iron strap hinge.

A rustic handcarved wooden hinge.

What hinge could be better for a horse barn than this?

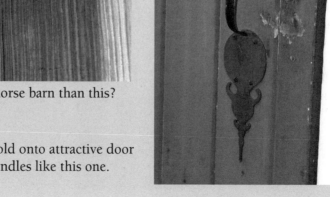

Hold onto attractive door handles like this one.

Utilities

Many outbuildings are perfectly utilitarian without any added "utilities" at all. Woodsheds and garden sheds, for example, often don't even need lights. An office, on the other hand, might need complete electrical wiring, full plumbing, heating and air conditioning, and several phone lines for voice, data, and fax. A full treatment of all utilities, even full treatment of any one, is far too complex for a single book on renovating outbuildings. The tips and advice that follow will deal with those aspects of electrical wiring and water supply that are unique to outbuildings, along with providing some information here and there that might be unfamiliar to someone who engages in wiring and plumbing only occasionally.

Electricity

The electrical wiring in an outbuilding suffers from many abuses that the wiring in a house seldom encounters. For example, outbuilding wiring may be chewed up by animals, either domestic or uninvited; it may get wet, either because the building is not well maintained or because it never was built to keep out blowing rain and snow; and it may receive physical abuse because often it is not protected by an inner finish wall. All of this rough treatment can easily result in power leaks.

Always remember that power leaks in an outbuilding can be far more dangerous than in your house. The reason for this greater danger is grounding. In your house you are usually standing on a suspended floor made of not very conductive materials and brushing up against not very conductive walls. In an outbuilding, on the other hand, you are more likely to be standing directly on earth, perhaps even damp earth, and brushing up against machines and fixtures that are highly conductive and well grounded. Under these circumstances, a power leak that might simply grab your attention in your house can take your life in an outbuilding.

So, my first bit of advice when working on the wiring in an outbuilding is: **If you are ever not absolutely sure that you are doing something correctly, don't do it at all.** Remember, this book is not a complete treatise on wiring. Consider, too, that if a bit of wiring is not important enough to be worth the expense of calling an electrician, then it's certainly not important enough to warrant the risk of making a dangerous mistake.

Outdated wiring

The first question you should raise if your outbuilding already has wiring is whether it is too old to be safe. This can be a very complex question, one that only a knowledgeable and widely experienced professional can answer in many situations. A few guidelines follow:

- If the insulation is brittle and easily rubs off, replace it.
- If the insulation consists of rubber and fabric (dating to before World War II), replace it.
- If you're unsure of its condition, but there isn't much wiring anyway, replace it.
- If you're unsure, but there's a lot of wiring that you'd like to use if you can, call a pro.

Inadequate lighting

If your existing wiring, including the branch circuit to the outbuilding, is in good condition but is being used to capacity, you don't necessarily need to bring in more power to improve the lighting level. You may well be better off by making more efficient use of the capacity that you already have. Three aspects of your lighting situation are worth looking at: the efficiency of the lamps you're using, the efficiency of the reflectors in the fixtures you're using, and the amount of light lost to light-absorbing surfaces.

Fluorescent lighting provides much more illumination per watt consumed than incandescent lamps. Even those odd-looking compact fluorescent bulbs that screw into fixtures designed for incandescent bulbs provide four times as much light per watt as the incandescent bulbs they replace. Their energy savings will pay for their greater cost even without considering the money you'll save by not having to run a new branch circuit to the outbuilding. They do take a bit of time, especially at low temperatures, to come up to full light output, so use them first in locations where you typically leave the light on for an extended period.

Reflectors are equally important in getting the most useful light out of limited capacity. Many outbuildings are equipped with simple bulb holders — no reflector at all. As much as half of the light generated may be shining into areas where you need no light. Replace the bulb holders with proper fixtures that incorporate good reflectors — and while you're at it, consider replacing them with fluorescent fixtures. Even if your fixtures do have reflectors, they may be rusted or so dirty that they don't reflect the light. Clean them or replace them.

The third area where you can improve lighting levels without adding capacity is the reflectivity of room surfaces. When light strikes a white or very light surface most of it is reflected back and contributes to the overall illumination of the space — it's recycled, so to speak. When walls, ceilings,

10-1. Barns often have few windows to let in light, so adding adequate interior lighting is frequently necessary.

If you're concerned about inadequate outdoor lighting, consider high-pressure sodium lights. They're expensive but put out an enormous amount of light for the power they consume. They also have a 24,000-hour life expectancy — that's nearly three years if you leave them on day in and day out, which I certainly hope you won't do. If you have to haul out an extension ladder to change your yard light, these fixtures may be worth the price even without their energy savings.

and floors are dark, they absorb light. A couple gallons of off-white paint will cost you a lot less money and effort than burying a new branch circuit cable and wiring new lighting circuits.

Even if you tackle all three of these areas and still consider the light levels inadequate, you haven't lost anything. All three of them will pay for themselves by saving electricity and by minimizing the amount of new lighting you will need to install.

Adequate wiring

Having decided whether existing wiring needs to be replaced, and whether you need to increase light output, you also need to decide how much power you are likely to require. Even if you are going to use the existing wiring, you may find that you need to add greater capacity.

The only safe way to ensure adequacy is with paper, pencil, and arithmetic. The object of the exercise is to ensure that the capacity of the wiring exceeds the loads that you will put on it. The box on the facing page gives the capacity of wires of various gauges in amps. Many electrical devices tell right on their nameplate how many amps they draw. Others, like light bulbs, are rated in watts. To convert, divide the wattage by the voltage (120 volts in the United States) to get the amps. For example, if your shop area needs five fixtures drawing 250 watts each, the total wattage is 1250 watts. Dividing the total wattage (1250) by the voltage (120) we find the total amperage to be 10.4 amps. From the chart we find that 14-ga. wire will handle this adequately. One final word on this arithmetic: Don't load circuits to more than 80 percent of their capacity.

10-2. With the difficulty of running underground lines to outbuildings it is essential to make sure that adequate wiring for all future uses is installed.

Wire sizes and types

Properly sizing your wiring is important for several reasons:

1. Wiring that is too small overheats and could cause a fire, or short circuit if it melts the insulation.

2. When wiring is too small and overheats, you are losing energy to the heating, and your electric bill goes up.

3. When wiring is too small, voltage drops. This can cause motors to overheat and burn out.

The table below provides the capacity (amperage) of readily available wire sizes. The smaller sizes (larger numbers) are typically used for wiring individual circuits, while the larger sizes (smaller numbers) are useful for running branch circuits, such as from your house to your outbuilding. For example, you might use 14 AWG (American Wire Gauge) cable for a 15 amp light circuit and 12 AWG for each of two 20 amp outlet circuits. If these three circuits (65 amp total) meet all of your needs, you could run the branch from your house with 4 AWG wire of proper type, assuming that the run is no more than 100 feet.

MAXIMUM CURRENT CAPACITY OF 100 FEET OF COPPER WIRE

Wire size	14	12	10	8	6	4	2
Max. amps	15	20	30	40	55	70	95

HOW-TO GUIDE

Rules of thumb

If you don't know exactly what your electrical loads will be you can use some rules of thumb. The first rule of thumb is for general lighting. This rule says to allow 3 watts per square foot of room area. Since we're only going to load a circuit to 80 percent of capacity, a 15-amp circuit will only be loaded to 12 amps. Converting to watts, 12 amps times 120 volts equals 1440 watts. At 3 watts per square foot, such a circuit will serve 480 square feet — that's 20 feet by 24 feet, a big room but a small barn.

The second rule of thumb is for outlets that will serve minor appliances. The typical household application is the kitchen, but the rule works well for shop spaces where portable power tools will be used. (You may think that your belt sander is much more powerful than your toaster — it's certainly louder — but that overlooks the fact that heating devices, like the toaster, draw heavily on current.) The rule of thumb says to provide two 20-amp circuits for a room that will be used by one or two people at a time. Having two circuits allows you to plug two heavy loads into separate circuits.

The third rule of thumb is to provide a separate circuit for each stationary piece of equipment.

However you work out your needs, if you expect that you will ever use power tools in your outbuilding, be sure to run at least two circuits to the building and keep the lights on a separate circuit from the outlets that will serve the power tools. If the tool happens to trip the circuit breaker you don't want to be left in the dark.

The proper wire or cable type is just as important as the proper size. Different types are appropriate for different applications. For example, nonmetallic cable for dry environments, such as is used in residences, is not suitable for a barn. If parts of your outbuilding could become wet you will need to use wire for a wet environment. The following are the types you are most likely to use.

Types of wire

DESCRIPTION	TYPE	USAGE
Armored (metal clad)	dry	For dry environments where physical protection is needed
Armored (metal clad)	wet	For wet environments where physical protection is needed
Nonmetallic	dry	For dry, protected environments
Nonmetallic	wet	For wet, protected environments
		For direct burial in soils that are totally free of stones, such as for a buried branch circuit
		For burial in a conduit where small stones may be present, such as for a buried branch circuit
		For an overhead branch circuit

Installing new electrical runs and lighting

The safe way to install wiring is to begin by installing all of the connection boxes and wires, then to make the connections beginning at the fixtures and other devices and working back toward the supply. By working in this sequence, nothing has been energized until you're done working on wires.

Getting down to the business of wiring lights and outlets, everything will be served by a "hot" (black or red) wire and a neutral (white) wire. When putting a switch in a circuit so you can turn it off, your safety dic-

TO BE SAFE

Just because "hot" wires are always dangerous, don't think of neutral wires as safe. A neutral can "bite" you for a variety of reasons.

◆ White wires are often used for the "hot" side of a circuit when wiring three- and four-way switches. While they should then be marked, not everyone is as careful as you should be, and the marking could be lost or cut off.

◆ Neutral wires carry current to and from the ground. If connections are poor or the wire is overloaded, *you* might be the path of least resistance to the ground.

◆ Things go wrong. While modern wiring practices as specified in the National Electrical Code are as safe as is practical, the materials and devices are made and installed by fallible human beings.

tates that you turn off the dangerous, "hot," black or red side of the circuit, not the neutral side.

NEVER connect a white wire to a black or red one. Later in this chapter you'll see some diagrams that require a cable carrying two "hot" lines, one to and one from a switch. Since cable doesn't normally come with two black or red wires, you'll have to use cable made with a black and a white wire. Color *both* ends of the white wire black or red *before* you connect them to anything. You can color them with a good permanent marker or by wrapping them with electrician's tape. Keep in mind that if the job is worth doing then it's worth doing so it will outlive you. Do the job right so that someone who outlives you also outlives working on the wiring that you did.

Continuing with the color coding, devices like outlets and light fixtures often have screw terminals where you connect your wires. They don't use black, red, or white screws but they come close. For red they use brass-colored screws, for white they use silver-colored screws. Always connect the black or red wire to the brass-colored screw and the white wire to the silver-colored screw.

Now we come to the weakest part of the color coding convention. For several decades wiring practice and virtually all wiring codes have required a separate grounding wire. This grounding wire is connected to any exposed conductive surfaces just in case something comes loose somewhere — you don't want the handle on your shop vacuum to unexpectedly carry 120 volts. This separate ground wire does not have its own insulation — it's bare, and copper color since it's copper. Do *not* connect it to the brass-colored screw on the side of an outlet. Connect it to the green screw — think "green ground."

These conventions are followed on devices like light fixtures, switches, and outlets, but the lugs for making connections in distribution boxes (load centers and the like) may be all silver in color. Follow the labels. Again, if you're not sure what you're doing, don't do it.

E X T R A   H E L P

Unless your place is wired for industrial-strength electrical equipment, your connection to the power grid consists of two "hot" wires and a neutral wire. Each of the "hot" wires alternates between a positive strength of 120 volts and a negative strength of 120 volts. When one of them is positive the other is negative, making the difference between them 240 volts.

When you wire an outlet or light with one of the hot wires and the neutral, you have wired it for 120 volts. That's the standard in this country for lights and outlets. If you run a circuit with two "hot" wires, one from each of the incoming

"hot" wires, you have wired it for 240 volts. Such a circuit might serve a stationary woodworking machine, or a well pump.

Certain conventions make it easy to keep track of what wires do what. The most important of these is the coloring of the insulation on wires. Neutral wires are white or occasionally gray. "Hot" wires are black or red. "Hot" wires are the most dangerous (unless, of course, somebody has wired things wrong or something has gone haywire.) If you think you might ever forget which is which, think of black as funeral black and red as blood red.

Burying a branch circuit

In my opinion, the choice between running a branch circuit over-head or underground is a no-brainer. No-brain, run overhead. With-brain, run underground. (I'll concede an exception if you're running a branch in a granite quarry.) The arguments are several:

Overhead lines are subject to damage from wind, ice, falling trees and branches, and tall vehicles, while underground lines are protected by tons of solid earth.

If an overhead line comes down you have a live wire endangering every living creature in the vicinity. If something should go wrong with an underground line it will most likely immediately trip a breaker or blow a fuse but in any event is still safely buried.

Overhead lines place a lot of stress on their anchor points, particu-larly in nasty weather. This requires that you keep an eye on them and may need to repair them from time to time. Underground lines don't stress anything.

Overhead lines are a nuisance when you periodically maintain your buildings. You shouldn't use an aluminum ladder in their vicinity, yet you will have to paint around them, usually from high up on a ladder. Underground lines are safely out of the way.

Last of all, overhead lines are ugly, while underground lines are out of sight.

If you're running a water line to your outbuilding, dig a single trench to accommodate both the water line and the branch circuit. Dig the trench to well below the frost line for the water line as described under the section on plumbing (page 235), then partially backfill, tamp-ing as you go, and lay in the branch circuit. Cover the branch circuit with a foot or so of earth and lay in a plastic warning tape. This will tell some future excavator that he's getting close if anyone should ever have to dig it all up, not a likely event if you do it right the first time. Then finish filling the trench. The proper wire to bury has more insulated conductors than you might suspect. Code, and therefore safety, requires that your total system have a single ground, usually quite close to where your power drop from the grid comes in. This means that your outbuilding branch should *not* have a separate ground rod. Instead, all of the grounding wires coming into your branch distribu-tion box should be connected to their own bus bar in the box, not the neutral bus bar. The grounding bus bar should then be connected, by an insulated conductor of the same gauge as the "hot" branch conductors, to the grounding bus at the service entrance panel. So, if you're running two "hot" lines to the outbuilding you'll need to bury four insulated conductors, two for the "hot" lines, one for neutral, and one for the grounds. (The same is true for an overhead line, though the over-head grounding line need not be insulated — it can be the bare "messenger" line that supports the insulated conductors — see the section on overhead lines below.)

In most parts of the country native soils have enough stone in them to require protection for a buried branch circuit. The code allows you to provide this protection by surrounding the wires with 6 inches to a foot of sand. As a practical and economic matter, however, you're usually better off running the wire in plastic conduit. You can then backfill with soil containing stones of up to about baseball size. The whole procedure is much simpler than I anticipated the first time I did it. The steps I follow:

1. Dig the trench, casting all of the dirt to one side and leaving the other side clear for assembling the line.

2. Lay out the cable (or separate wires) alongside the trench.

3. Thread one end of the wires through a length of conduit and slide the conduit along the wires to about their midpoint.

4. Thread the wires through another length of conduit, slide it along to the first, and cement the joint.

5. Continue from both ends until the straight conduit is fully assembled and cemented.

6. Nudge one end into the trench and watch as the whole thing flops down. There's a surprising amount of flexibility in the assembled conduit.

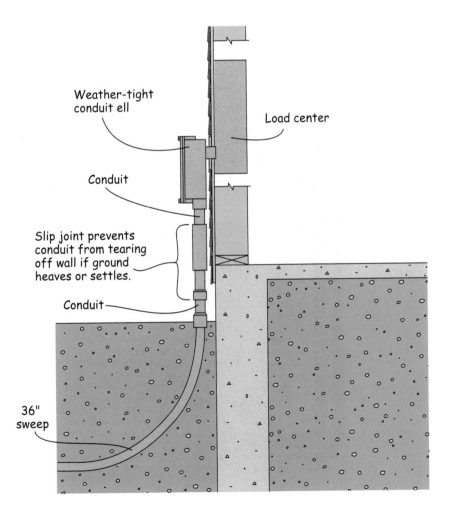

10-3. This service entrance shows a typical connection between a load center and an underground line.

I've found that even 2-inch conduit will easily follow any curve that a backhoe can dig as a continuous trench.

7. Once it's in the trench, thread the sweeps onto the wires and cement them in place, bringing your branch up to ground level.

8. Backfill by hand, avoiding large rocks, until the conduit is covered by several inches of soil, then let a machine finish the job. Don't forget the warning tape a foot or so above the conduit.

Running an overhead branch

To run an overhead branch (if you insist) you need to use a cable consisting of insulated conductors spiraled around a bare "messenger." The messenger carries the suspended weight and can also serve as the grounding wire. In most cases you will need three insulated conductors plus the messenger, two insulated "hot" wires, an insulated neutral, and the bare ground. You must leave the insulated wires free of tension. Any overhead wiring must be at least 10 feet above the ground, at least 15 feet above any driveway, and at least 18 feet above any public road. Local ordinances may specify greater clearances. Depending on the length of the run, the anchor points at the two ends will need to be considerably higher than this, of course, because you can't pull the cable into a straight line. You also need to allow for expansion and contraction as tempera-

10-4. This small load center may be all you need for an outbuilding. It can provide two protected circuits using conventional breakers or four using a tandem breaker.

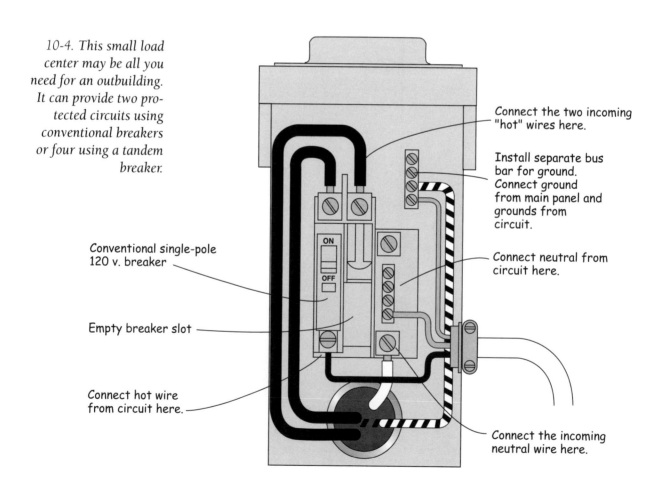

Connect the two incoming "hot" wires here.

Install separate bus bar for ground. Connect ground from main panel and grounds from circuit.

Connect neutral from circuit here.

Conventional single-pole 120 v. breaker

Empty breaker slot

Connect hot wire from circuit here.

Connect the incoming neutral wire here.

tures fluctuate. The wires need to maintain ground clearance during the hottest temperatures of the year without becoming as taut as violin strings during the coldest temperatures. Get the advice of an electrician or the power company's local line superintendent. Anchor the ends of the messenger to a building or mast of adequate strength with a forged eyebolt. You can attach the messenger to the eyebolt by working an eye-splice in the end of the messenger or by using a special cable support grip.

Installing the branch box

Leaving the source end of your branch circuit entirely unconnected for the time being, route the outbuilding end to a suitable load center.

Plan the location carefully. The simplest and easiest connection consists of a load center on the inside of the wall with a conduit ell connecting to its back from outside the wall. The ell connects to the buried conduit via a slip joint that allows for some heaving or settling of the soil without ripping everything from the wall. You can paint these above ground conduit connections if you like, but I suggest that you wait a year or so. If the backfill settles a bit, withdrawing the inner part of the slip joint from the outer part, you'll be left with an unpainted band around it.

10-5. An example of a larger-capacity load center.

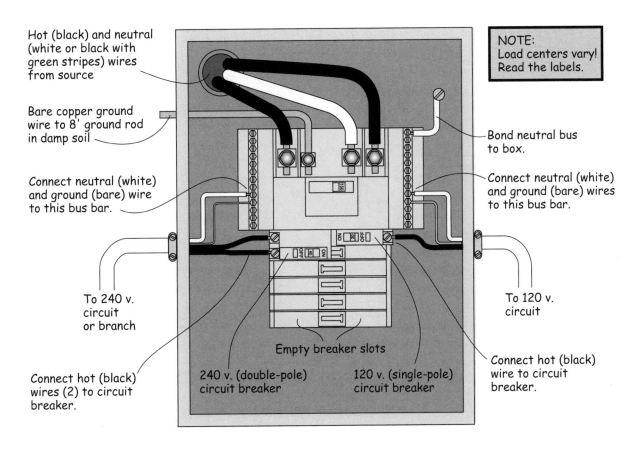

Hot (black) and neutral (white or black with green stripes) wires from source

Bare copper ground wire to 8' ground rod in damp soil

Connect neutral (white) and ground (bare) wire to this bus bar.

NOTE:
Load centers vary!
Read the labels.

Bond neutral bus to box.

Connect neutral (white) and ground (bare) wires to this bus bar.

To 240 v. circuit or branch

To 120 v. circuit

Connect hot (black) wires (2) to circuit breaker.

240 v. (double-pole) circuit breaker

Empty breaker slots

120 v. (single-pole) circuit breaker

Connect hot (black) wire to circuit breaker.

Wiring switches and outlets

The drawings that follow are a reminder of how to connect these most common devices. The wire lengths shown are strictly for clarity in the drawing — none of your actual wires should be less than 6 inches long, but they should not be much more than that without good reason. Note how the ground wires are handled in each case.

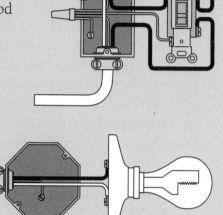

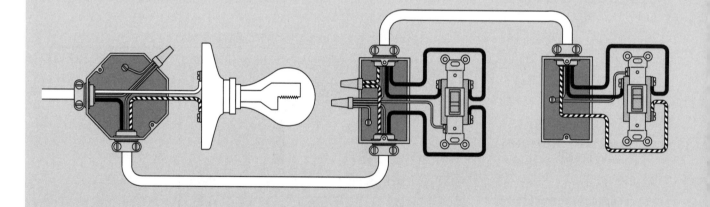

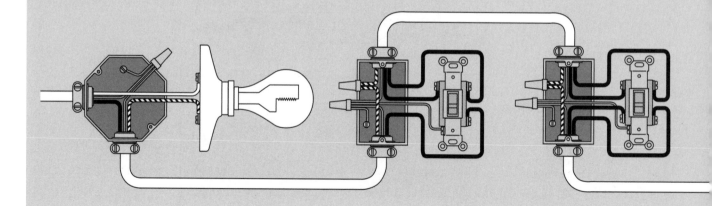

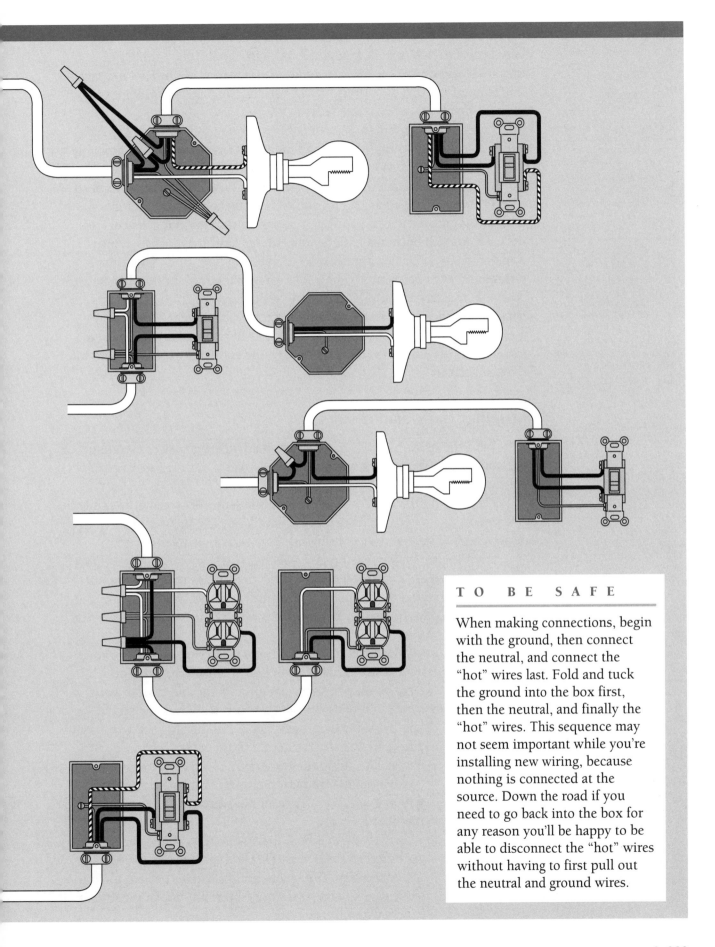

TO BE SAFE

When making connections, begin with the ground, then connect the neutral, and connect the "hot" wires last. Fold and tuck the ground into the box first, then the neutral, and finally the "hot" wires. This sequence may not seem important while you're installing new wiring, because nothing is connected at the source. Down the road if you need to go back into the box for any reason you'll be happy to be able to disconnect the "hot" wires without having to first pull out the neutral and ground wires.

Choose a load center, sometimes called a breaker box or fuse box, based on the total current that you're bringing in and the number of breaker slots you will need. The small box shown in *Figure 10-4* on page 230 is fine for many situations. It will handle 70 amps and provides two slots for breakers.

Larger load centers such as the one in *Figure 10-5* on page 231 have a main breaker in addition to individual circuit breakers. This is probably more like the one from which you will be running the branch circuit.

Note an important difference in the way these two boxes have been wired: The small box has had an additional bus bar installed for all of the grounding wires. It is connected to a ground wire that goes back to the main service entrance panel and to the ground rod in that vicinity. The neutral bus bar has not been bonded to the box itself. This is how you should wire your branch box. The larger box on the other hand has both the neutrals and the grounding wires connected to the same bus bars and the bus bars have been bonded to the box itself. This is how your service entrance panel should be wired. If you install a larger box for your outbuilding branch, wire the grounding wires as shown here for the smaller box.

Installing individual circuits

With the branch box in place you're ready to run individual circuits. For any but the simplest of wiring schemes, plan it out on paper before you begin.

Keep the number of connections in a single box as small as possible. You can do this by wiring from box to box like beads on a necklace rather than by wiring several outlets or lights back to a single junction box.

Don't be stingy with the boxes — use the biggest ones that will fit in the wall. A small box with too many connections crammed inside is not just a frustration and a nuisance, it's dangerous trouble in the making and a violation of code. It's usually easiest to install all of the boxes before you run the wire.

When running the wire, follow the route that offers the most protection from possible abuse. This is particularly important in an outbuilding where activities may be more rough-and-tumble than in your living room and where you may not plan a finish wall over the stud spaces. Whenever possible, avoid draping the wire from a hole in one stud to a hole in the next — choose routes like along the underside of a plate. When you do pass through a framing member, keep the hole centered where it will be safer from nails driven into the framing member later on.

Arriving at a box, make sure the cable is securely anchored to the box and leave a generous 6 inches inside the box. You want enough wire to be able to conveniently make a solid connection but not so much that excess wire fills the box.

TO BE SAFE

In damp or wet environments or where floors are earth or concrete over earth, use only ground fault outlets. Code allows ground fault circuit breakers instead of outlets and they are certainly just as safe but you'll probably find it less expensive to use the ground fault outlets.

Once all of the boxes and wire are in place, install the fixtures, switches, and outlets and make the necessary connections inside the boxes. Double-check every connection as you go, giving each individual wire a tug to make sure it's secure. If you don't, you're sure to have wires come loose as you fold them into the box, or, worse, after everything is closed up.

The final step in the outbuilding is connecting the circuits at the load center. First connect the grounding wires to the ground bus bar, then the neutrals to the neutral bus bar, and finally connect the "hot" wires to a circuit breaker and plug in the breaker. You should leave the breaker off for the time being.

Back at the main entry load center, connect up the outbuilding branch circuit using a breaker of proper size for the total capacity of the outbuilding wiring and turn it on. Then, back at the outbuilding, turn on the breakers one at a time. Assuming that there are no shorts and the breaker stays on, check that all of the outlets, switches, and lights work properly.

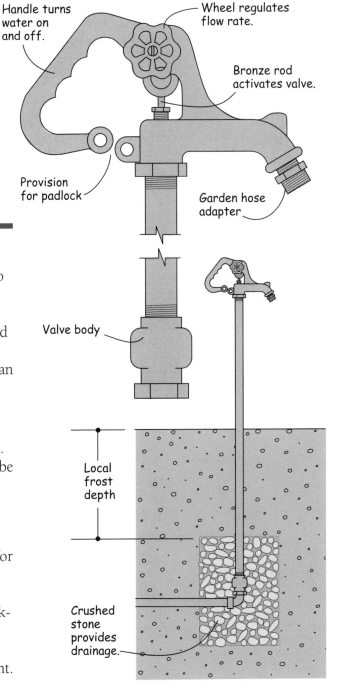

10-6. The valve of a yard hydrant operates below frost level. Water in the standpipe drains out when the hydrant is off.

Water

In most parts of North America, water supply to an outbuilding is in danger of freezing. While frozen pipes may be a minor and temporary inconvenience in some cases, if the freeze is hard and the pipes or fixtures burst because of the pressure exerted by the expanding ice, then it can become a substantial expense. The suggestions that follow cover three common needs:

◆ Water on demand during all seasons.
◆ Water on demand only during warm seasons.
◆ Water supply to a heated building that may be shut down for extended periods.

Four-season water

Barns and stables require drinking water for the animals regardless of temperatures. While farm supply stores, either local or mail-order, provide a variety of heaters to keep actual drinking troughs open, supplying water to them is a quite different problem. The traditional and proven solution is a device called a yard hydrant.

A yard hydrant is supplied by a water line buried below the local frost depth. A valve buried at that same depth controls the flow of water to a standpipe that extends above ground. When the valve is closed a drain is simultaneously opened, allowing water in the standpipe to drain back to below the frost line. As long as the soil around the valve allows the drain water to percolate into the ground, the hydrant supplies running water regardless of temperature. (*See Figure 10-6.*)

You can install a single yard hydrant or more than one in a series by connecting to the main water line with an ell or a tee as appropriate. Keep in mind, too, that you don't have to install a yard hydrant in a yard — if you can bring the main line into your outbuilding below the frost line then you will be able to have the convenience of water on demand inside your barn or stable.

Yard hydrants are manufactured as assembled units for specific burial depths ranging from 1 to 8 feet to accommodate the anticipated depth of frost in various parts of the country. They are designed so that interior parts can be withdrawn from the top for replacement if needed without having to dig up the valve. Use only bronze ells or tees, not plastic, so minor shifting doesn't cause leaks that would require digging to repair. Use two stainless steel clamps to make the below ground connections.

Proper drainage is essential for a yard hydrant. Surrounding the valve with crushed stone is essential but not sufficient if you've dug down into a stratum of solid clay or other non-draining soil. Percolation from aboveground through the disturbed material will quickly fill the spaces between the stones and the hydrant won't drain. The only way to make it work is to bury a drain tile to daylight downhill.

One final precaution: If you connect a garden hose to the hydrant, you must either remove the hose after each use or install a vacuum breaker between the hose and the hydrant. The hydrant won't drain if air can't enter the standpipe.

Warm-season water

If your outbuilding requires water only during the frost-free months of the year you have more options.

If a single source will meet your needs, burying the supply below the frost line is not a problem, and you don't want the seasonal task of draining the system, then use a yard hydrant as just described.

If you want the most economical system and the terrain between your source of supply and your outbuilding permits burying a water line with a continuous downhill grade in one direction or the other, then bury a water line a foot or so deep regardless of frost depth and drain it before danger of frost. Just make sure it will fully drain. This can be very difficult when using black plastic pipe if the slope of the line is very slight — any short stretch where the pipe slopes in the wrong direction can retain enough water to split the pipe in a freeze. Install a shutoff to isolate the outbuilding supply line, install a boiler drain at the lowest level of the line to drain the system, and if the supply end is the high end, install a boiler drain there too to admit air while draining. To drain the system, shut off the supply and open both ends of the line.

If the terrain prevents you from installing a water line that will drain properly, you have no choice but to bury the line below frost. At the outbuilding end, while still below the frost line, install a curb stop. This is a shutoff connected to a standpipe to ground level and capped at ground level. You often see the caps in sidewalks in urban areas. You'll also need to buy one of the special handles to insert down the standpipe to turn the water on or off. The best design is called the Minneapolis pattern — it provides a threaded connection between the valve and the standpipe, thus ensuring proper alignment. Be sure to get the type called "stop and drain" and install it so that the downstream end drains. Proper drainage is as important for the proper functioning of a stop and drain curb stop as it is for a yard hydrant, so see the precautions discussed above.

Water for occasionally heated buildings

Water supply to a building that is heated spring and fall but not during the coldest months is similar in many ways to water supply for warm months only — as long as the supply is below the frost line the rest of the plumbing is out of danger until the heat is turned off. The difference is that such buildings are likely to have more extensive plumbing. The suggestions that follow address the problem of making it as easy as possible to fully drain the system before turning off the heat.

The first concern is protecting the water supply. Install a curb stop as described above under "Warm-season water."

The second concern is protecting the distribution. All of the pipes must be easily and thoroughly drained. This requires that you pitch all of the plumbing much as you normally would for waste water and sewage systems. As a practical matter, this means that you totally avoid flexible plumbing such as plastic pipe and copper tubing — use only rigid copper and support it at close enough intervals that it doesn't sag between supports when full of water. Plan the layout carefully, keeping in mind that you can't drop down to go around an obstruction. At the lowest point in the supply system, tee down to a boiler drain. If you do it right, shutting down for the winter should be as simple as turning off the water at the curb stop, opening all of the taps, and then draining it all through the boiler drain.

One other issue should be mentioned here: the proper disposal of waste water. While you don't need a complete septic system for wash water, you shouldn't just dump it out in the grass either. The equivalent of a small leach field built along the lines of a leach field for a septic system will serve for small amounts of wash water — just don't go dumping nonbiodegradables into it. If you put a conventional trap in the drain of a sink leading to such a leach field, be sure to fill it with nontoxic antifreeze before turning off the heat.

Metric Conversion Chart

WHEN THE MEASUREMENT GIVEN IS	TO CONVERT IT TO	MULTIPLY IT BY
inches	centimeters	2.54
feet	meters	0.305
square feet	square meters	0.0929
pounds	kilograms	0.45
pounds per square foot	kilograms per square meter	4.88
miles	kilometers	1.6
°F	°C	°F − 32 x $\frac{5}{9}$

Index

Note: Page numbers in *italics* indicate illustrations or photographs.

Other Storey Titles You Will Enjoy

Be Your Own House Contractor, by Carl Heldmann.
The book to help you save 25 percent on building your own home —
without lifting a hammer!
176 pages. Paper. ISBN 978-1-58017-840-2.

Build a Classic Timber-Framed House, by Jack A. Sobon.
Complete instructions and plans for building a hall-and-parlor home.
208 pages. Paper. ISBN 978-0-88266-841-3.

Building Small Barns, Sheds & Shelters, by Monte Burch.
Comprehensive coverage on tools, materials, foundations, framing,
sheathing, wiring, plumbing, and finish work for outbuildings.
248 pages. Paper. ISBN 978-0-88266-245-9.

Home Plan Doctor, by Larry W. Garnett.
The essential companion to buying home plans, from understanding
basic design principles to requesting needed modifications.
224 pages. Paper. ISBN 978-1-58017-698-9.

How to Build Animal Housing, by Carol Ekarius.
An all-inclusive guide to building shelters that meet animals' individual needs: barns, windbreaks,
and shade structures, plus watering systems, feeders, chutes, stanchions, and more.
272 pages. Paper. ISBN 978-1-58017-527-2.

How to Build Small Barns & Outbuildings, by Monte Burch.
Complete plans and instructions for more than 20 projects, including an add-on garage,
a home office, a roadside stand, equipment sheds, and four types of barns.
288 pages. Paper. ISBN 978-0-88266-773-7.

Rustic Retreats: A Build-It-Yourself Guide, by David and Jeanie Stiles.
Illustrated, step-by-step instructions for more than 20 low-cost, sturdy,
beautiful outdoor structures.
160 pages. Paper. ISBN 978-1-58017-035-2.

Storey's Basic Country Skills, by John and Martha Storey.
A treasure chest of information on building, gardening, animal raising, and
homesteading — perfect for anyone who wants to become more self-reliant.
576 pages. Paper. ISBN 978-1-58017-202-8.

Timber Frame Construction, by Jack Sobon and Roger Shroeder.
Clear explanations of the basics of timber frame construction.
206 pages. Paper. ISBN 978-0-88266-365-4.

These and other books from Storey Publishing are available
wherever quality books are sold or by calling 1-800-441-5700.
Visit us at *www.storey.com*.